Andrew Wheen

# Dot-Dash to Dot.Com

## How Modern Telecommunications Evolved from the Telegraph to the Internet

 Springer

Published in association with
Praxis Publishing
Chichester, UK

Dr. Andrew Wheen
Baldock
Herts
UK

SPRINGER–PRAXIS BOOKS IN POPULAR SCIENCE
SUBJECT *ADVISORY EDITOR*: Stephen Webb, B.Sc., Ph.D.

ISBN 978-1-4419-6759-6          e-ISBN 978-1-4419-6760-2
DOI 10.1007/978-1-4419-6760-2
Springer New York Dordrecht Heidelberg London

Library of Congress Control Number: 2010929000

Cover design: Jim Wilkie
Project copy editor: Christine Cressy
Typesetting: BookEns, Royston, Herts., UK

Printed on acid-free paper

Springer is part of Springer Science+Business Media (www.springer.com)

# Dot-Dash to Dot.Com

How Modern Telecommunications Evolved from the Telegraph to the Internet

# Contents

*For Carol, Laura and Alex*

# Acknowledgments

A large number of people have contributed to this book in one way or another, and I am grateful to them all. However, particular thanks are due to the following:

Eric Benedict, who first suggested that there might be a need for this book.

Clive Horwood and the Praxis staff, for guiding me through the intricacies of the publication process.

Stephen Webb, for reviewing the manuscript and for contributing many helpful suggestions.

Robert Dudley, Francis Wheen and Julia Jones, for the benefit of their extensive knowledge of the publishing industry.

David Brown, John Davies, Andy Doyle, Bob Partridge, David Posner and Joe Savage, for acting as referees.

John Jenkins, Sam Hallas, Professor Nigel Linge, Keith Schneider, Don Johnson and Motorola Heritage Services, for permission to use their photographs.

Wikimedia Commons, for making a wonderful range of old photographs available on the Internet.

Colleagues at Mott MacDonald and ex-colleagues at Mentor, for many thought-provoking conversations.

And, finally, to my family and friends, for their dependable support and encouragement during the years that it has taken me to write this book.

# Figures

# Pictures

# Tables

# About the author

Andrew Wheen has worked in the telecommunications field since 1982. He has held senior engineering and product management roles with major suppliers of telecommunications equipment and was one of the original architects of the Energis network in the United Kingdom (now part of Cable & Wireless). More recently, he has worked as a management consultant in the telecommunications and broadcasting industries. Dr Wheen is a Fellow of the Institution of Engineering & Technology and is a Chartered Engineer. He is married with two children and lives near London.

# Introduction

Methods of communicating over long distances advanced surprisingly little from the days of the Roman Empire to the start of the nineteenth century. Of course, beacons had been lit to warn of the arrival of the Spanish Armada in the English Channel in 1588[1] and semaphore flags had been used to pass messages between ships at sea (such as Nelson's famous "England expects ..." signal before the battle of Trafalgar). However, the speed at which information could be transmitted was typically limited by the speed at which a horse could gallop or a ship could sail. As a result, it took 2 months for news of Nelson's sensational victory at the Battle of the Nile to reach London in 1798.

The lack of fast and reliable communications could have devastating consequences. In 1812, a message was sent by sea from England to inform the United States that the British would no longer interfere with American shipping. Unfortunately, the Unites States declared war over the issue while the message was still crossing the Atlantic. Fighting had already broken out by the time that the message arrived and nearly 4,000 British and American soldiers were killed before hostilities ended. If better communications had been available, the war of 1812 might never have happened.

The lack of high-speed, long-distance communications wasn't just a problem for government and the military—it was a major impediment in many different areas of economic and social activity. At the start of the nineteenth century, newspapers were publishing foreign news that could be months out of date, while businesses struggled to communicate with salesmen and customers located just a few hundred miles away. Communication with friends and family in different parts of the country was subject to the vagaries of a painfully slow and unreliable postal system.

Against this background, the appearance of the first practical telegraph system in the 1830s can be seen as a defining moment in the history of telecommunications. Here, at last, was a method of communicating over long distances with minimal time delay. The word "telecommunication" literally means "sharing over long distances"[2] and the telegraph would eventually expand to cover the whole globe. It is the invention of the telegraph—and its dramatic consequences—that provides the starting point for this book.

The telegraph paved the way for telephone networks. These, in turn, laid the foundations for the Internet. In little more than 170 years, simple arrangements of wire and mechanical switches evolved to become the largest and most complex machine in the world. It is a machine that spans every continent and reaches down into some of the wildest and most inhospitable places on Earth. It is a machine that is still developing rapidly in new and unexpected directions and is delivering services as diverse as television, social networking, podcasting and the World Wide Web. It has become an essential part of life in the twenty-first century.

This book covers some of the major events in the history of telecommunications and explains how today's digital technology has evolved from devices that were no more complicated than an electric doorbell. The aim is to appeal to a wide audience by treating the subject with a light touch and by not assuming any pre-existing expertise on the part of the reader. For this reason, non-essential technical details and background information have been confined to appendices and notes at the end of the book.

# 1 The birth of an industry

On the evening of March 3, 1843, Professor Samuel Morse and his friend, Mr Henry Ellsworth, were standing in the lobby of the United States Senate. There was tension in the air. It was the last day of the Congressional session, and any Bills that had not been signed into law by midnight would be lost.

Morse had known failure before. After he demonstrated the operation of his electric telegraph to President Van Buren in 1838, a Bill had been introduced into Congress to provide $30,000 to build an experimental telegraph system between Washington and Baltimore. The Bill had been favorably reviewed by the commerce committee, but had made no further progress. Similar Bills had been introduced in each succeeding session of Congress, but had been equally unsuccessful.[1]

A friendly senator approached Morse and informed him regretfully that there were 119 Bills ahead of his in the queue for voting. This meant that the chances of the Bill being put to a vote that evening had effectively disappeared. Morse was so discouraged that he left the Capitol, bought a train ticket to New York City for the following morning, and returned to his boarding house to pack. As he retired to bed that night, he resolved to waste no more time on the electric telegraph.

But Henry Ellsworth was not so easily defeated. He remained where he was, and continued to lobby for the Bill. Finally, with 5 minutes to go before the Senate adjournment, the Bill was passed without a vote. The President signed the Bill into law moments before the midnight deadline.

The next morning, Ellsworth sent his 17-year-old daughter, Annie, round to Morse's boarding house to tell him the good news. She arrived while he was having breakfast. In the excitement of the moment, Morse promised Annie that she would be the first to send a message over the new telegraph system when construction was complete.

Today, most people have largely forgotten the electric telegraph. However, it can still be seen in Western movies when the clerk is desperately tapping out a message to warn that the Indians are coming. From the creation of the first practical telegraph systems in the 1830s until the invention of the telephone in the 1870s, the telegraph was the only available method of sending messages at high speed over long distances. It continued to be widely used until well into the twentieth century, and its impact on the Victorian world was every bit as significant as the Internet has been in more recent times.

\* \* \*

A. Wheen, *From Dot-Dash to Dot.com: How Modern Telecommunications Evolved from the Telegraph to the Internet*, Springer Praxis Books, DOI 10.1007/978-1-4419-6760-2_2,
© Springer Science+Business Media, LLC 2011

Samuel Morse was an unlikely inventor. He had been born in 1791 in Charleston, Massachusetts, and had shown an early interest in drawing. In 1805, he entered Yale College to study chemistry and natural philosophy, and it was there that he received his first instruction in electricity from Professor Jeremiah Day. In 1809, he wrote that:

> "Mr Day's lectures are very interesting; they are upon electricity; he has given us some very fine experiments, the whole class, taking hold of hands, form the circuit of communication, and we all received a shock apparently at the same moment. I never took an electric shock before; it felt as if some person had struck me a slight blow across the arms."

However, the young Morse showed more of an aptitude for art than for science, and he eventually wrote to his parents informing them that he had decided to become a painter:

> "My price is five dollars for a miniature on ivory, and I have engaged three or four at that price. My price for profiles is one dollar, and everybody is willing to engage me at that price."

In 1811, he moved to London, where he exhibited at the Royal Academy and achieved considerable success. Returning to America in 1815, he established a studio in Boston, from where he subsequently moved to New York. By 1829, he was back in Europe, studying in Paris and in Italy. Here, he remained until 1832, when he returned to New York.

By 1837, Morse had become disillusioned with art. He had been forced into portraiture in order to earn a living, but many of his paintings had been of questionable quality. Furthermore, he had been deeply disappointed by his failure to win an expected commission to paint a mural for the Rotunda in the Capitol building in Washington. It was time to try something completely different. ...

During his voyage back to America in 1832, Morse had struck up a conversation with a fellow passenger, Dr Charles T. Jackson. Jackson had been studying electricity and magnetism in Europe, and their conversation covered many recent developments in the field. It appears that Morse was captivated by the idea that electricity could be used to transmit information at very high speed over long distances.[2] The remainder of his voyage was spent drawing up plans for an electric telegraph.

Morse's telegraph was based upon the use of an electromagnet[3] and a battery. In concept, it was no more complicated than a door bell where the pushbutton and the bell are separated by a very long length of wire. Although Figure 1 does not correspond exactly with the system developed by Morse, it helps to illustrate the principles involved.

A Morse telegraph key is simply a switch. When the telegraph key is closed, current flows from the battery to energize the electromagnet, and this causes the armature to move downwards. When the key is opened, the electromagnet is de-

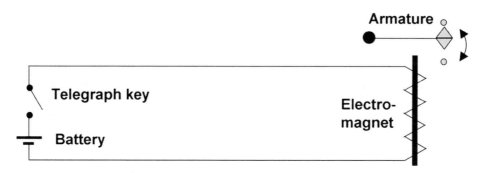

**Figure 1.**  Basic telegraph.

energized and the armature springs back to its former position. As a result, the position of the armature at the receiver always indicates the setting of the telegraph key at the transmitter.

This, clearly, has the potential to allow communication at a distance, but it is not sufficient as it stands. If the operator at the receiving end notices that the armature is moving back and forth, it tells him that someone is fiddling about with the key at the transmitter, but it does not convey much useful information. In order to convert the telegraph from a communication device that can convey just one message (such as a doorbell or a fire alarm) to a communication device that can convey any required message, some form of code is required.

Morse recognized this problem. His solution has become known as the Morse Code and it was an integral part of his design for the electric telegraph. Morse Code represents each letter of the alphabet as a series of dots and dashes, as shown in Table 1.[4]

**Table 1.**  Morse Code.

| | | | | | |
|---|---|---|---|---|---|
| A | · - | N | - · | 0 | - - - - - |
| B | - · · · | O | - - - | 1 | · - - - - |
| C | - · - · | P | · - - · | 2 | · · - - - |
| D | - · · | Q | - - · - | 3 | · · · - - |
| E | · | R | · - · | 4 | · · · · - |
| F | · · - · | S | · · · | 5 | · · · · · |
| G | - - · | T | - | 6 | - · · · · |
| H | · · · · | U | · · - | 7 | - - · · · |
| I | · · | V | · · · - | 8 | - - - · · |
| J | · - - - | W | · - - | 9 | - - - - · |
| K | - · - | X | - · · - | . | · · - · · · |
| L | · - · · | Y | - · - - | , | - - · · - - |
| M | - - | Z | - - · · | ? | · · - - · · |

As can be seen, some letters (such as E) require fewer dots and dashes than others (such as Z). The aim here is to assign the shortest codes to the letters that occur most frequently in English, thereby reducing the length of the coded message. Morse determined the letter frequencies by counting the number of copies of each letter in a box of printer's type.

Although Morse Code can be transmitted at any speed that suits the operators at the transmitter and receiver, the relative timing of the dots, dashes and gaps is critical. Clearly, a dot must be significantly shorter than a dash if the two are to be correctly distinguished at the receiver. It must also be possible to differentiate between similar dot/dash combinations such as J (· - - -) and A (· -) M (- -). For these reasons, the rules state that:

- a dash should be equal to three dots;
- the space between parts of the same letter should be equal to one dot;
- the space between two letters should be equal to three dots;
- the space between two words should be equal to five dots.

Not surprisingly, telegraph operators required a significant amount of training.

The development of Morse Code was driven by the needs of the telegraph, but it turned out to have a much wider range of uses. Morse Code was adopted for radio communication in the 1890s and was used as an international standard for maritime communication as late as 1999. Morse Code is still used by amateur radio enthusiasts and there have even been suggestions that it might be integrated into mobile phones to supplement SMS text capabilities. The telegraph may have died, but Morse Code lives on!

In 1832, Morse was appointed Professor of Painting and Sculpture at the University of the City of New York—a post that seems to have provided him with plenty of time to pursue his experiments with telegraphy. In 1835, he built a working prototype and the following 2 years were spent making a series of improvements to the equipment. By September 1837, Morse was able to demonstrate his telegraph operating over 1,700 feet of wire laid out in coils across a room. He immediately filed for a US patent. Three months later, he made his formal request to Congress for the funds to build an experimental telegraph between Baltimore and Washington.

The primary challenge faced by Morse once the funding for his experimental telegraph had been approved was to install the cable. The original plan had been to encase the cable in lead and bury it underground using a specially designed plough. The two conductors would be given an insulating coating by wrapping them in cotton and then dipping them in a bath of hot gum shellac. However, the distance between Baltimore and Washington is approximately 40 miles, and it was found after 10 miles of cable had been buried that the insulation had been damaged by the process of encasing the cable in lead. This was a major crisis. Twenty-three thousand dollars of the government's $30,000 appropriation had been spent, and the installed cable was completely useless.

Many around him felt that the situation was hopeless, but Morse proved equal to the challenge. Recognizing the need to keep his problems out of the newspapers, he arranged for the plough to "accidentally" hit a protruding rock and suffer serious damage. The newspapers carried sensational accounts of the incident, suggesting that it would take several weeks to complete the repairs. However, the gentlemen of the press failed to detect the much bigger problem that lurked, quite literally, beneath their feet.

After some consideration, Morse concluded that the only viable option would be to dispense with the lead casing and mount the cable on poles. Overhead wires could be mounted far enough apart to avoid the need for an insulating coating and glass insulators could be used to prevent any electrical leakage through the poles. This form of construction had originally been rejected because it was feared that the cable would be vulnerable to vandalism, but the pressures of the situation—coupled with the news that overhead wires had been used successfully in England—convinced Morse that it was the only acceptable solution. To this day, we refer to poles carrying telephone lines as "telegraph poles".

The line was finally completed in May 1844. Morse remembered his promise to Annie Ellsworth and asked her if she had chosen the inaugural message. She replied that she and her mother had selected a biblical quotation: "What God hath wrought!"[5] On May 24, 1844, Morse successfully transmitted this message from the Supreme Court chamber in Washington to the Mount Clare depot in Baltimore.

For many people, this was the event that marked the birth of the telecommunications revolution. However, a public telegraph service had started operating in London 5 years earlier. As is so often the case with pivotal moments of this type, the invention of the telegraph was the sum of the accumulated efforts of a number of talented people over many years. Without detracting from Morse's extraordinary achievements, it must be recognized that he had the good fortune to be working in the field at the moment when these individual contributions coalesced into a viable new technology.

\* \* \*

The need to communicate rapidly over long distances has existed for as long as human beings have been around. However, for most of recorded history, the available techniques have been rather unsatisfactory. Messengers on horseback can travel long distances, but it can still take a considerable

Samuel Morse.

time for the message to arrive at its destination, and the system is inherently unreliable and insecure. Systems based upon beacons or semaphore flags are much faster but are useless in fog or heavy rain.

The idea of using electricity to send messages down a wire has a surprisingly long history. As early as 1753, a letter appeared in the *Scots' Magazine* under the heading "An Expeditious Method of Conveying Intelligence". This letter is believed to have been written by a Scottish inventor called Charles Morrison, although it was signed simply "C.M.". Morrison proposed—but never built—an electrostatic telegraph system in which 26 insulated wires conducting static electricity from a Leyden jar[6] would be able to cause movements in 26 small pieces of paper carrying the letters of the alphabet.

However, static electricity creates very high voltages. These voltages would have been difficult to handle with the cable insulation materials that were available at the time, so the requirement to use 26 separate wires—each adequately insulated from all the rest—would have presented major technical challenges. To address this problem, a number of alternative systems requiring fewer wires were proposed over the next half-century.

In 1816, Francis Ronalds demonstrated an electrostatic telegraph based upon Morrison's original concept, but using just one wire. A friction-based electrostatic machine was used to charge the line and pith ball electroscopes detected the presence of charge at the receiver. The system is illustrated in Figure 2.

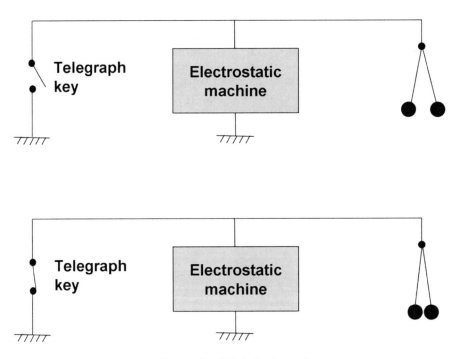

**Figure 2.** Pith ball telegraph.

The pith ball electroscope consists of two small balls hanging in contact with each other on pieces of thread. If the telegraph key is opened, then the balls will be charged by the electrostatic machine; since like charges repel, the balls will fly apart. When the telegraph key is closed, the balls are discharged and so will no longer repel each other. In other words, the spacing of the pith balls at the receiver provides a direct indication of the setting of the telegraph key at the transmitter.

Although this arrangement is sufficient to tell an operator at the receiver that there is someone operating the telegraph key at the transmitter, the problem of how to transmit an intelligible message from one end to the other has not yet been solved. Which letter of the alphabet is the sender trying to convey with each closure of the telegraph key? Morrison's approach had been, in effect, to build 26 parallel telegraphs—one for each letter of the alphabet—but Ronalds had already decided that his system would operate over a single wire. Ronalds's solution to the problem required a clock at each end of the link. The second hand of each clock was replaced by a circular disk carrying the letters of the alphabet written around its edge. A mask with a small window in it was placed in front of the disk so that only one letter was visible at a time. As the disk rotated, a sequence of letters was displayed in the window. If the clocks at the transmitter and the receiver were synchronized, then they would display the same letter at any given moment. Once this had been achieved, a message could be transmitted by momentarily opening the telegraph key whenever the next required letter appeared in the window.

Ronalds demonstrated a working telegraph in Hammersmith in 1816—a remarkable achievement for the time. However, with the benefit of hindsight, we can see that the system had a number of weaknesses. To begin with, it was slow. After each letter was transmitted, the sender needed to wait until the next required letter appeared in the window before he could transmit again. If the required letter had only just passed the window, then there could be a significant delay. This delay could be reduced by increasing the speed of rotation of the disk, but there comes a point at which mistakes start to be made because the receiving operator cannot determine whether the required letter is the one just leaving the window or the one just appearing. This problem would become worse if the two clocks were running at slightly different speeds; whilst ways could be found to align the clocks at the start of a message, they would gradually drift out of alignment. The solution to these problems arrived less than 20 years later with the invention of the Morse Code. Although Morse Code could only be used by trained operators, it proved to be a faster and more reliable way of communicating.

A further problem with Ronalds's telegraph was its limited range. Ronalds was able to operate his system over a distance of 8 miles, but it is not clear how a telegraph based on electrostatic principles could have been extended to cover much longer distances. Morse's use of batteries rather than an electrostatic machine proved to be critically important in freeing the telegraph from the constraints of distance and opening up the way for the development of a global communications system.

\*  \*  \*

Seven years before Morse's telegraph transmitted its first message between Washington and Baltimore, William Fothergill Cooke and Charles Wheatstone had successfully transmitted and received a message over a telegraph line running along the railway track between Euston and Camden Town in North London. In 1839, the world's first public telegraph service had opened on a 13-mile route between London Paddington and West Drayton, providing members of the public with the opportunity to send messages for 1 shilling a time.

Cooke and Wheatstone's telegraph was based on a discovery in 1820 by the Danish physicist, Hans Christian Oersted, that an electric current carried by a wire can cause a compass needle to deflect. A horizontal row of five iron needles was arranged across the middle of the diamond, as illustrated in Figure 3. These needles hung in a vertical position when no current was flowing in the wires.

The Cooke and Wheatstone telegraph. The arrangement within the diamond-shaped part of the apparatus is illustrated in Figure 3.

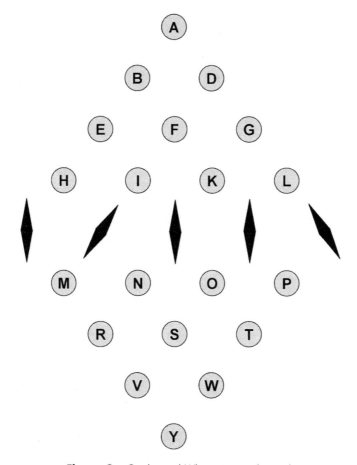

**Figure 3.** Cooke and Wheatstone telegraph.

However, by pressing the appropriate combination of switches at the transmitter, a pair of needles could be deflected so that they pointed to one of the 20 letters displayed on the diamond. In the case shown in Figure 3, the letter D has been selected.

By adopting this design, Cooke and Weatstone had restricted themselves to 20 of the 26 letters of the alphabet. The letters that they chose to omit were C, J, Q, U, X and Z, so words such as "quiz" would have presented quite a challenge! However, the telegraph operators soon found ways around this problem. A classic example occurred on January 1, 1845, when the telegraph played a crucial role in the capture of a murderer.

On that day, Mr John Tawell traveled to Slough with the intention of poisoning his mistress. It appears that he had expected her to die quietly and was greatly disconcerted when she let out a bloodcurdling scream. To make good his escape, he dashed out into the street and ran to the train station. There, as luck would have it, he found a train that was just about to depart for London. As he climbed aboard and settled into his seat, he was firmly convinced that he would have disappeared into the London crowds long before news of his crime reached the city.

However, he had not reckoned on Cooke and Wheatstone's telegraph, which had been installed along the railway line between Slough and Paddington. The following message was flashed up the line:

> "A MURDER HAS JUST BEEN COMMITTED AT SALT HILL AND THE SUSPECTED MURDERER WAS SEEN TO TAKE A FIRST CLASS TICKET TO LONDON BY THE TRAIN WHICH LEFT SLOUGH AT 742PM HE IS IN THE GARB OF A KWAKER WITH A GREAT COAT ON WHICH REACHES NEARLY DOWN TO HIS FEET HE IS IN THE LAST COMPARTMENT OF THE SECOND CLASS COMPARTMENT."

The word "KWAKER" caused some confusion at Paddington until it was realized that the intended word was "QUAKER"—the telegraph was not capable of transmitting Q or U. When Tawell arrived in London, the police were waiting for him. He was subsequently found guilty of murder and hanged.

This episode provided confirmation—if any were needed—of the value of the telegraph. However, in addition to its inability to transmit six letters of the alphabet (or any punctuation characters), Cooke and Wheatstone's telegraph had a number of other weaknesses relative to Morse's design. To begin with, their telegraph required six separate wires to link the transmitter to the receiver. Although the number of wires was eventually reduced to one by replacing the five-needle telegraph with a system using just one needle, this change made the system much more complex to operate, so it could no longer be used by unskilled operators. Furthermore, the operator was required to manually record each message as it arrived at the receiver because the power behind a deflected needle was too small to punch holes in a paper tape or to drive the pen for a recording device. By using an electromagnet in his receiver, Morse was able to deliver sufficient power to provide a permanent record of each message.

\* \* \*

The problem of signal attenuation on long cables is one that is still faced by telecommunications engineers today. Although the electrical resistance of a short piece of copper wire is negligible, the resistance starts to become significant if the length of the wire is increased. A strong transmitted signal will become steadily weaker as it travels down the wire. If nothing is done to boost this signal, then it may no longer be intelligible by the time it arrives at the receiver.

Morse was well aware of this problem and he solved it by using a device called a "relay". A relay is essentially a switch controlled by an electromagnet. When an electric current flows in the electromagnet, it attracts a strip of metal and thereby causes the switch to close. When the current stops flowing in the electromagnet, the strip of metal is no longer attracted to it and the switch springs open again.

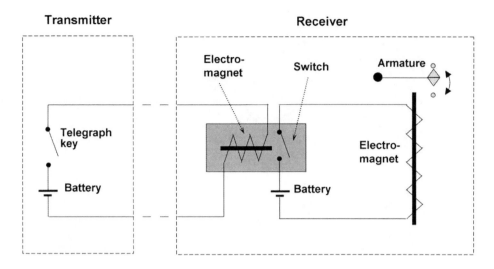

**Figure 4.** Telegraph relay.

If a relay is placed at the end of a long cable, it can be used to increase the power of received signals. The principle is illustrated in Figure 4, with the relay shown highlighted.

When the telegraph key is closed at the transmitter, current flows in the circuit and the electromagnet in the relay is activated. This causes the switch in the relay to close, thereby causing a second electromagnet in the receiver to be activated. This, in turn, causes an armature to move and create an audible sound. The armature has to be significantly heavier than the switch in the relay and so a larger current is required in the second electromagnet. Notice, however, that the power required to move the armature is not coming from the transmitter—it is being supplied by a second battery located at the receiver. In effect, the relay is enabling the switch at the transmitter to control the delivery of power to the receiver, but *without the need for that power to be transmitted down the line.*

This elegant solution has the additional benefit that the relay can be used to control another relay further down the line. Very long telegraph lines can be built by placing batteries at regular intervals and using relay-based devices called repeaters to regenerate the signal. By 1861, the telegraph system spanned the United States from the Atlantic to the Pacific. A trans-Atlantic link was established 5 years later and by 1872, there was a link to Australia. The telegraph was going global.

\* \* \*

Messages sent by telegraph were known as "telegrams".[7] To send a telegram, it was normally necessary to go to the local telegraph office where the message and the postal address of the destination would be written on a form. This form would then be handed to the telegraph operator—along with the appropriate

Telegraph sounder and key. A device designed for audible reception of Morse Code is known as a "sounder". The photograph shows a sounder (left) and a telegraph key (right) dating from the 1860s.

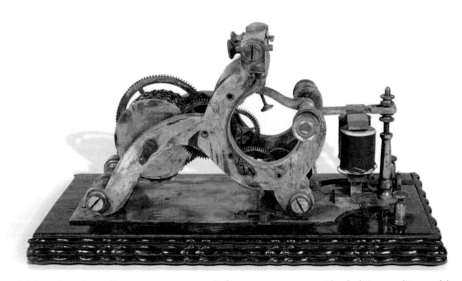

Telegraph register. If a permanent record of a message was required, the sounder would be replaced by a device called a "register". A number of different forms of register were devised, but they typically used a paper tape that was drawn through the machine by rollers driven by a clockwork motor or a descending weight. An electromagnet was used to bring a metal stylus into contact with the paper tape when current was flowing in the telegraph circuit, thereby causing some form of mark to be left on the paper.[8] The paper moved at a constant speed, so the length of the mark corresponded to the duration of each received dot or dash. The photograph shows a very early weight-driven register dating from 1851.

fee—and the message would be transmitted down the line. At the receiving telegraph office, the armature in the telegraph receiver would create an audible *click-clack* noise corresponding to the dots and dashes in the incoming message. The receiving operator would decode the Morse Code characters as he heard them and would write or type the received message on a blank telegram form. Large numbers of young boys were employed to deliver incoming telegrams to their intended recipients and to bring back any replies to the telegraph office for transmission.

*   *   *

The development of the telegraph in the middle of the nineteenth century was every bit as significant as the development of the Internet at the end of the twentieth century. A magazine article published in the United States in 1873 gives a rather breathless account of the range of applications for the new invention:

> "Every phase of the mental activity of the country is more or less represented in this great system. The fluctuations in the markets; the price of stocks; the premium on gold; the starting of railroad trains; the sailing of ships; the arrival of passengers; orders for merchandise and manufactures of every kind; bargains offered and bargains closed; sermons, lectures and political speeches; fires, sickness and death; weather reports; the approach of the grasshopper and the weevil; the transmission of money; the congratulations of friends—every thing, from the announcement of a new planet down to an enquiry for a lost carpet-bag, has its turn in passing the wires."[9]

The telegraph was adopted most rapidly in the United States, where the installed length of telegraph cable grew from 40 miles in 1846 (Morse's Washington-to-Baltimore link) to over 12,000 miles by 1850. In Great Britain, there were 2,215 miles of cable installed by 1850 and a number of other countries also had rapidly growing telegraph networks. Even the invention of the telephone did not stop the expansion of the telegraph network—telephones were relatively rare in private homes until after the Second World War,[10] so telegrams were used for rapid communication with non-subscribers.[11]

Many new jobs were created by the technology, while some existing occupations were rendered obsolete. The widespread adoption of Morse Code created a demand for skilled telegraph operators. Telegraphers often had to endure long shifts and unpleasant working conditions, and the repetitive pounding of the telegraph key gave rise to one of the earliest forms of industrial injury (known as Glass Arm or Telegrapher's Paralysis). In spite of these drawbacks, telegraphy was seen as an attractive profession and telegraphers became respected members of the community. For the ambitious, telegraphy provided an escape route from small towns to the big cities; for the restless, telegraphy skills meant guaranteed work wherever they went. The first signs that skilled telegraph operators might eventually become an endangered species did not occur until the 1920s, when teleprinters started to become widespread. These machines could transmit from an ordinary keyboard and so did not require operators with a working knowledge of Morse Code.

It was the need to control the operation of railway networks that drove the expansion of the telegraph network beyond the major cities to smaller towns. The early telegraph pioneers were quick to establish relationships with railway companies and these relationships proved to be highly symbiotic; the telegraph operators gained the rights to install their cables along railway lines (thereby saving the trouble of having to conduct negotiations with a large number of separate landowners), while the railway company gained access to telegraph services (which proved to be critically important to the running of an efficient railway network). Morse's experimental link from Washington to Baltimore was installed along a railway line, as was Cooke and Wheatstone's initial system between Paddington and West Drayton in London.[12]

The British government soon found that the telegraph was an essential tool for running an empire that spanned the globe. In India, the telegraph and the railway enabled the British Raj to maintain control over a huge country with just 40,000 troops. The development of submarine cables meant that it was possible to run cables directly from Britain to outposts of the empire without having to rely on the goodwill of countries along the route. As a result, control of the empire could be centralized in London without having to worry about the security of the communications channels.

One of the earliest military applications of the telegraph occurred in 1844 when Cooke and Wheatstone's system was used to provide a direct link from the Admiralty in London to the naval base at Portsmouth. Telegraph systems were subsequently used in military campaigns during the Crimean War, the American Civil War and the Boer War. A number of military victories have been directly attributed to the effective use of telegraph systems on the battlefield.

Surprisingly, the development of the telegraph was initially viewed as a serious threat by the newspaper industry. Many felt that newspapers would never be able to compete with the telegraph for delivery of up-to-the-minute news and so would have to focus on retrospective analysis and commentary. However, far from killing off newspapers, the telegraph actually opened up new opportunities for the industry. In an era in which news could take weeks to arrive, the telegraph suddenly enabled newspapers to report events that had happened on the previous day. "Time itself is telegraphed out of existence," boasted the *Daily Telegraph*, a newspaper whose name had been chosen to give the impression of rapid delivery of news. The telegraph was still being used by newspapers as late as the 1930s and telegraphers were regularly seen at sporting events, natural disasters, war zones and other newsworthy places.

However, many newspapers could not afford the expense of maintaining reporters in the far-flung corners of the world that had now been opened up by the telegraph. Furthermore, telegraph companies charged by the word, making it expensive to send back long dispatches. The solution to these problems was for newspapers to club together to share the costs, leading to the development of newspaper wire services. Paul Julius von Reuter had established an agency in Europe during the 1840s to distribute financial information by carrier pigeon. However, he soon recognized the potential of the telegraph and when France and England were linked by cable in 1851, he moved his headquarters to London. Although his reports were initially aimed at business customers, he soon started selling dispatches to newspapers. In America, the New York Associated Press began offering similar services to newspapers in 1848.

For most of the nineteenth century, people had to rely upon the accuracy of sundials to set their clocks and watches, and so towns and cities that were just a few miles apart often operated on different local time. This was not a problem while the available forms of travel were all so slow, but the arrival of the railway created the need for a single standard time. In 1883, Western Union launched a time service in conjunction with the US Naval Observatory. On certain telegraph wires, normal service was suspended just before noon and the word TIME was repeated over and over again in Morse Code. At midday, a single dot was sent down the wires, thereby enabling clocks across the United States to be synchronized.

Today, it is hard for us to recognize just how different the telegraph was from anything that had gone before it. Although professionals were quick to grasp the potential benefits of the telegraph for business and commerce, many ordinary individuals found it hard to understand how communication was possible without the physical transfer of a piece of paper. One elderly lady refused to accept a telegram from her son because she was sure that it was not his handwriting on the form. People would hand in messages to be transmitted and then watch the telegraph wires to see if they could see their message as it was sent on its way. Once it became possible to transfer money by telegraph from one place to another, people believed that notes and coins were being physically transmitted and they started turning up at telegraph offices with other small objects that they wished to send by telegraph.

Although telegraph operators were allowed to swap stories of this kind, they were definitely not allowed to discuss the contents of any of the messages that they sent or received. The Rules, Regulations and Instructions of the Western Union Telegraph Company, published in 1866, make this quite clear:

> "All messages whatsoever—including Press Reports, are strictly private and confidential, and must thus be treated by employees of this Company. Information must in no case be given to persons not clearly entitled to receive it, concerning any message passed or designed to pass over the wires or through the offices of this Company."

This rule applied even when messages appeared to be related to illegal activities and discretion was certainly required when transmitting messages of a personal nature. Some business organizations sent their commercially sensitive messages in code so that they would be meaningless to the telegraph operators who handled them.

# 2   The telegraph goes global

By the early 1850s, the telegraph network was expanding rapidly and there was a growing need to lay cables across stretches of water. In view of the problems that had been encountered with insulation on land-based cables, it was clear that underwater cables would present some major challenges. In 1842, Morse installed a submarine cable between Castle Garden and Governor's Island in New York Harbor, but the cable was damaged by a ship's anchor after transmitting just one message. In 1845, Ezra Cornell set up a 12-mile underwater cable between Fort Lee, New Jersey and New York City. This cable had worked for several months before it was broken by an ice flow. In 1850, a 25-mile underwater cable was installed between England and France, but it was cut by a French fisherman after only 3 days.

Despite these problems, a cable was successfully laid between England and France in 1851. It remained in service for many years, proving conclusively that seawater was no longer a barrier to the expansion of the telegraph network. By the end of the 1850s, a large number of submarine cables had been installed in both Europe and North America, but these early installations were really just setting the scene for the *real* challenge—a cable across the Atlantic.

The successful laying of a trans-Atlantic cable was one of the most extraordinary feats of engineering in the history of telecommunications. The hero of our story is Cyrus Field, an American businessman who had made his fortune in the New York textile and paper trade. At a time when existing submarine cables rarely exceeded 100 miles in length, Field was planning a submarine cable 2,000 miles long[1] laid between Ireland and Newfoundland, with an additional 1,000 miles of cable to link Newfoundland with New York City.

Field joined up with four other New York businessmen to form the grandly named New York, Newfoundland and London Telegraph Company. They decided to start with the "easy" part of the project—the link between Newfoundland and New York City. Although this was mainly an overland route, the difficult terrain and the appalling weather presented considerable difficulties. Furthermore, the first attempt to lay an 85-mile submarine cable across the Cabot Strait was a complete disaster and they had to return for a second (successful) attempt the following year. The Newfoundland-to-New-York-City link was finally completed in 1856.

At this point, the company had exhausted its funds, so Field turned to England for additional investment. In October 1856, he created the Atlantic Telegraph Company and started selling stock in London. One of Field's great strengths was a complete inability to accept defeat. The following list of his

A. Wheen, *From Dot-Dash to Dot.com: How Modern Telecommunications Evolved from the Telegraph to the Internet*, Springer Praxis Books, DOI 10.1007/978-1-4419-6760-2_3,
© Springer Science+Business Media, LLC 2011

attempts to lay a trans-Atlantic cable stands as testimony to his extraordinary persistence:

*Attempt No. 1*
In 1857, *USS Niagara* and *HMS Agamemnon* were each loaded with 1,250 miles of cable and they set sail from Southern Ireland. The plan was that the *Niagara* would lay its half of the cable first and the *Agamemnon* would then take over when they reached mid-Atlantic. However, the cable broke after only 440 miles had been laid.

*Attempt No. 2*
For the second attempt, it was agreed that the *Niagara* and *Agamemnon* would both start laying cable in mid-Atlantic and would proceed in opposite directions. However, on their way to the starting point, the ships encountered a massive storm that lasted for 6 days and threatened to capsize both vessels. Despite 45 crew members requiring treatment for injuries, they did eventually manage to start laying cable, but the cable snapped almost immediately. They started again and this time they managed to lay 80 miles of cable before it snapped. On the third try, the cable snapped after 220 miles had been laid.

*Attempt No. 3*
On July 28, 1858, *Niagara* and *Agamemnon* set sail once again and this time they managed to complete the installation of the cable. However, attempts to communicate over the installed cable were less successful—it was found that the received signal was so weak that over half the words needed to be repeated. In an attempt to rectify the problem, the company's Chief Engineer increased the voltage. To his consternation, this burnt out the cable's insulation and rendered it useless. The public rejoicing that had followed the first successful laying of a trans-Atlantic cable was brought to an abrupt and acrimonious end.

*Attempt No. 4*
After this string of failures, it was decided that the next attempt would have to use a cable that was almost three times heavier. This raised the question of how many ships would be needed to install such a massive cable. The problem came to the attention of the famous engineer, Isambard Kingdom Brunel, who invited Cyrus Field to visit the construction site of the *Great Eastern*. At 693 feet long, this ship was five times the size of any other vessel afloat when it was completed in 1857, and it was capable of carrying the full length of cable needed to span the Atlantic. However, the first attempt using the *Great Eastern* met with failure when the cable broke just 600 miles short of the Newfoundland coast. Interestingly, the sudden loss of connectivity in the cable was detected back in Ireland, leading to speculation that the *Great Eastern* had sunk; there was considerable relief when it eventually returned to port.

*Attempt No. 5*
In July 1866, the *Great Eastern* left Ireland for another attempt and Field's

persistence finally paid off—after just 2 weeks, the cable had been successfully installed. Telegraph messages were soon being exchanged across the Atlantic and the link was opened up for commercial traffic. Amazingly, this cable was still in use 100 years later.

Later that summer, the cable that had broken during Attempt No. 4 was successfully repaired and the remaining 600 miles of cable were laid. Within a month of the first trans-Atlantic cable becoming operational, a second cable was also transmitting messages.

It is satisfying to note that this tale of dogged determination had a happy ending. The long sequence of failures caused widespread skepticism about the project and many investors decided to cut their losses before success was achieved. Western Union, the dominant telegraph operator in the United States, were so convinced that a trans-Atlantic cable would never work that they started building a telegraph link to Europe via Siberia—a route that required only 25 miles of submarine cable to cross the Bering Straits. These plans were scrapped when they heard of Field's achievement.

*  *  *

As we have seen, the early telegraph pioneers encountered major problems with cables. Since the cost of installing and maintaining cables was considerable, there was a strong commercial incentive to reduce the number of cables required. One way of doing this was to utilize the conductivity of the Earth itself to form one half of a telegraph circuit. This technique, known as an "earth return", is illustrated in Figure 5.

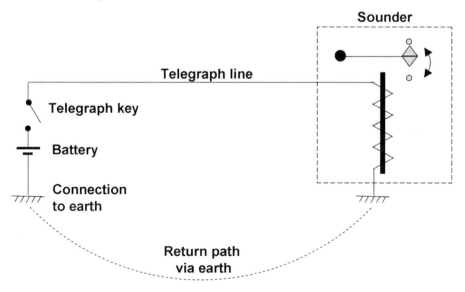

**Figure 5.** Earth return.

A good electrical connection to earth is required at each end of the line and these are then used to provide the return path for the telegraph circuit. A suitable connection to earth can be achieved by burying a metal object in a damp patch of ground or by immersing it in a nearby river or stream.

The earth return had been discovered in 1837 by Professor K. A. Steinheil of Munich University. Professor Steinheil had been investigating a suggestion by the famous mathematician, Carl Friedrich Gauss, that the two rails of a railway line could be used instead of wires to carry a telegraph circuit. Although it turned out that the insulation between the two rails was insufficient for this purpose, the experiment led to the accidental discovery that the Earth itself could be used as a conductor. Steinheil declined to patent this important discovery, dedicating it instead to the benefit of the public. It is interesting to note that Morse's Washington-to-Baltimore link (constructed 7 years later) initially used two metallic conductors, but an earth return was added before the public trial took place.[2]

The idea of an earth return is often surprising to people who have not encountered it before. We tend to expect electrical conductors to be made from metals such as copper or aluminum, and the surface of the Earth is not noticeably metallic. Certainly, if a copper wire was replaced with an equivalent volume of wet soil, the result would probably turn out to be a rather poor conductor of electricity. However, it must be remembered that the Earth is an extremely large lump of matter, so the return path is made up of a huge number of separate electrical paths. Since these paths are connected in parallel, the combined effect produces a return path of relatively low resistance.

The use of an earth return means that a single wire is sufficient to carry a telegraph circuit between two locations. However, on busy routes between large cities, a number of separate telegraph circuits may be required to handle the volume of traffic, thereby creating a requirement for a large number of cables. In order to reduce the cost of this cabling, inventors looked for ways of making one wire carry more than one telegraph circuit. From the earliest days of telegraphy, it had been possible to alternately send and receive traffic down a single wire, but the invention of the "duplex telegraph" enabled a single wire to carry traffic in both directions *simultaneously*. A description of one of the methods used to achieve this apparently impossible feat is provided in Appendix A.

The first duplex telegraph was invented by Dr Julius Wilhelm Gintl in Austria in 1853, but the technology was not commercially successful. It was subsequently improved by Carl Frischen in Germany. Even the great Thomas Edison worked on the design of a duplex telegraph and was greatly disappointed when he was beaten to the Patent Office by Joseph B. Stearns in 1872. However, Edison was not put off by this early setback and in 1874, he invented a quadruplex telegraph that could handle two simultaneous transmissions in each direction. He later went on to invent a sextuplex telegraph (three simultaneous messages in each direction) in 1891.

At about the same time as Edison was patenting his quadruplex telegraph,

Baudot distributor.

Emile Baudot in France was developing a sextuplex telegraph that worked on an entirely different principle. Baudot's system was based on a device called a "distributor", which was originally invented by Bernard Meyer in 1871.

In concept, Baudot's distributor is rather similar to the distributor used in some petrol engines. The various telegraph transmitters sharing the line were connected to electrical contacts that were arranged around the circumference of a circular drum. A shaft, driven either by weights or by an electric motor, rotated in the center of the drum. Metal brushes were mounted on this shaft in such a way that they touched each of the electrical contacts as the shaft rotated, thereby connecting the telegraph line in turn to each of the telegraph transmitters. At the other end of the line, a second distributor was used to distribute the incoming signal to each of the receivers. Electrical signals were sent down the line to keep the two distributors in step, thereby ensuring that the signals from each transmitter were delivered to the correct receiver. It is a technique that modern engineers would refer to as "Time Division Multiplexing".

For reasons that will soon become apparent, the Baudot system did not use Morse Code. The characteristics of the new code (which became known as Baudot Code) meant that a single telegraph key was no longer sufficient to transmit a message—instead, each operator required a device that looked like a very small piano, with five separate keys.

To transmit a letter, the operator held down the appropriate combination of keys. Once the keys had been pressed, they were not released until an audible

Baudot keyboard.

click had been heard to indicate that the distributor had transmitted the character to the far end of the line. It was then up to the operator to type in the next character before the distributor came round again. Baudot operators were typically required to transmit at a steady rate of two or three letters per second—something that required considerable skill.

Baudot did not adopt Morse Code for his new telegraph because complex codes such as "?" (..-..) take considerably longer to transmit than simpler codes such as "E" (.), and this does not suit a distributor-based system in which each operator is given a fixed amount of time to transmit. Using the Baudot Code, the code for each character is the same length, so the time taken to transmit each character is constant. The Baudot Code is described in Appendix B.

Baudot's multiplexing system had the significant advantage that it could, in principle, be enhanced to carry more channels.[3] If the circumference of the drum could be increased, then space would be created to accommodate additional electrical contacts, thereby allowing additional operators to share the same telegraph line. If the rotational speed of the shaft remained the same, then each operator would be able to continue transmitting at the same rate as before, but the number of characters per second carried by the line would increase. Of course, mechanical systems such as Baudot's are likely to run into practical difficulties at relatively low speeds, but the same technique can now be implemented using digital electronics. Modern networks transmit data at speeds of 10 Gbit/second (10,000,000,000 bit/second) and above over fiber optic cables using the same Time Division Multiplexing concept as Baudot used in 1874.

*   *   *

So far, our search for ways to improve the efficiency of a telegraph line has focused on increasing the number of telegraph circuits that it carries. However, another way of increasing the traffic-carrying capacity of a line is to increase the rate at which the information is transmitted by each separate telegraph circuit. Electrical signals travel down a telegraph wire at the speed of light, but the rate at which an operator can tap out his message is limited to a few thousand words per hour. This observation led to the development of the automatic telegraph, which can transmit pre-prepared messages down a telegraph line without the need for human intervention, and so can operate at very high speeds.

In 1846, a Scottish clockmaker named Alexander Bain patented a "chemical telegraph", which greatly increased the speed of telegraph transmission. Bain's

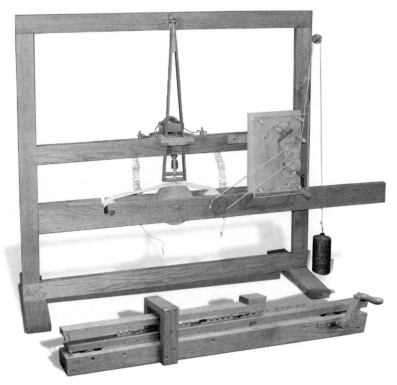

Reproduction of Morse's original design. Ironically, Samuel Morse's original experiments were based on a form of automatic telegraph. His original transmitter[4] used a wooden arm dragged across sawtooth-shaped pieces of metal to open and close an electrical circuit, with the arrangement of the teeth determining the content of the message. At the receiver, a pendulum with a pencil attached to its base rested on a moving paper tape; every time the electrical circuit was closed, an electromagnet would pull the pendulum to one side, and this deflection would be marked on the tape. By the time the Baltimore-to-Washington link was built, Morse had replaced his automatic transmitter with the (much simpler) telegraph key. The photograph shows a modern reproduction of Morse's original design.

system was based upon the use of a paper tape in which holes had been punched to represent the message. At the transmitter, the tape was read by using electrical contacts to detect the presence or absence of holes. At the receiver, the message was reproduced by using electrical impulses to create discolorations on a chemically treated paper tape. This method of receiving messages required few moving parts and so could operate at high speed. The chemical telegraph was demonstrated on a link between Paris and Lille, and managed to transmit 282 words in 52 seconds. At a time at which standard Morse systems were only achieving about 40 words per minute, this represented a considerable advance. Further development showed that the system was capable of operating at well over 1,000 words per minute.

Automatic telegraphs turned out to be very effective for transmitting long messages such as news reports, but it was often more economical to transmit shorter messages by hand. Bain's chemical telegraph demonstrated that paper tape was an effective method of storing messages prior to transmission over a high-speed telegraph, and it was widely used on many later telegraph systems. Indeed, paper tape was still being used as recently as the 1980s as a convenient method of storing computer data.

*  *  *

As we have seen, many nineteenth-century telegraph systems could only be used effectively by skilled operators. This ultimately proved to be a major weakness of the telegraph because the telephone—when it finally arrived—could be used by anyone. There were, however, a number of successful attempts to make the telegraph more user-friendly and these developments helped to extend the life of the telegraph network well into the twentieth century. We will now review some of these inventions.

As early as 1846, an American with the wonderful name of Royal House invented the first "printing telegraph", which could record messages on a strip of paper. Unlike the Morse register, which simply printed a sequence of dots and dashes, the printing telegraph produced its output in written English that could be read directly by the recipient. However, its use of electromagnets fell foul of Morse's patents and an alternative pneumatic version had to be patented in 1852.

House's machine, like many of those that followed, used a rotating disk to print the text at the receiver. The letters of the alphabet were distributed evenly around the edge of the disk and pulses were sent down the line to rotate the disk in fixed increments until the correct position was reached. A striker was then used to press the selected character against the paper. This machine, although complex, did give satisfactory results and it inspired further developments.

One of these developments led to the "Stock Ticker",[5] which was the forerunner of modern business information services. It was used to deliver market information such as stock prices to brokers and dealers in city-center offices. The Stock Ticker evolved from the "gold indicator", which had been invented in 1866 by Dr S. S. Laws. Dr Laws was the presiding officer of the Gold Exchange in New York and he developed the system to avoid the need for messenger boys running up and down Wall Street to deliver the latest gold price to the New York Stock Exchange. It soon became clear that the system could drive more than one receiver, thereby creating the opportunity to offer a subscription service to nearby offices. The service was expanded to cover a wide range of prices (not just gold) and the Stock Ticker was born. Laws's original design was improved by Callahan in 1867 and by Edison in 1869.

The Stock Ticker could receive information, but it could not transmit a response—it was optimized for a very specific, low-volume application. If the telegraph was to compete with the telephone, then businesses and other organizations had to be able to send and receive messages from their premises without the need for a messenger service to the local telegraph office. Furthermore, the necessity for such users to employ skilled Morse Code operators had to be avoided. A general-purpose terminal device was required that could transmit using a standard typewriter keyboard and could print out received messages as ordinary text.

These requirements led eventually to the development of the teleprinter. The first teleprinter[6] was developed Charles L. Krum and his son, Howard. In 1910, they installed a teleprinter link between New York and Boston and by 1914, teleprinters were being used by the Associated Press to deliver copy to newspaper offices throughout America. During the 1920s, a global teleprinter network known as the telex network was established. Despite the Internet, the telex network was still in use for certain forms of business communication at the start of the twenty-first century.[7]

The teleprinter undoubtedly made the telegraph network much more accessible to new types of users, but it could only transmit text. To transmit any form of photograph or diagram, a facsimile telegraph was required. Most people regard the fax machine as a relatively recent invention, but this is not in fact the case. As early as 1843—the same year as Morse was lobbying Congress for funds to build his telegraph link from Baltimore to Washington—Alexander Bain had patented an "electrochemical telegraph" that exhibited many of the characteristics of a modern fax machine. (We encountered Alexander Bain earlier in this chapter as the inventor of the chemical telegraph. Confusingly, the chemical telegraph and the electrochemical telegraph were two completely separate inventions.)

A fax machine can be used to transmit a black-and-white image across a network. The transmitting machine scans the image from side to side, starting at the top of the page and working down to the bottom. At each point during the scan, the transmitter sends an electrical signal that indicates whether that part of the image is black or white. The scanner in the receiving fax machine is synchronized with the scanner at the transmitter. If, at a particular point in the scan, the signal received from the transmitter says that the image is black, then the receiver will make a black mark on the paper. If, on the other hand, the signal indicates that the image is white at that point, then no mark will be made.

Clearly, such a signal could be transmitted down a telegraph line by adopting the convention that current flowing on the line indicates a black part of the image while a lack of current indicates white. However, building a facsimile transmitter and receiver with the technology available in the 1840s required some truly heroic engineering.

Bain used a pendulum at the transmitter to provide an even scan across the image, with the image being moved forward by a small amount after each swing of the pendulum. The image to be transmitted was copied to a sheet of copper and the white areas were etched away, leaving the black areas of the image standing proud of the surface. As the pendulum swung across the copper sheet, it completed an electrical circuit whenever it made electrical contact with one of these raised areas, and an electrical signal was sent down the line to the receiver.

At the receiver, an identical pendulum scanned across a sheet of paper soaked in potassium iodide, with the end of the pendulum making electrical contact with the paper. Whenever current flowed in the circuit, it caused a brown mark to appear on the paper. Since electromagnets[8] were used to synchronize the transmit and receive pendulums, the position of each brown mark at the receiver corresponded with a raised area in the copper plate at the transmitter. In this way, a facsimile of the original image could be transmitted down the line.

During the following years, a number of improvements were made to Bain's original design. In 1851, an English physicist named Frederick Bakewell demonstrated a "copying telegraph" at the World Exhibition at Crystal Palace in London. Bakewell's machine used a drum to hold the image instead of the flat-bed arrangement used by Bain. The image was drawn on tin foil using non-conducting ink, so a flow of current down the wire indicated a white area of the image.[9] Both machines used a scanning stylus that was in physical contact with the image at the transmitter and both used damp electrolytic paper to record the image at the receiver.

The first commercial facsimile services were established in France in 1865 by an Italian physics professor named Giovanni Caselli. The service linked Paris with several other French cities using a "pantelegraphe"—a cast iron contraption that stood more than 2 meters tall. "Kopiertelegraphen" facsimile machines by Bernhard Meyer were also introduced on the French telegraph service at about the same time. Over the next few years, the French facsimile service was gradually expanded. It was found that facsimile offered two significant advantages over the conventional telegraph: first, it virtually eliminated the possibility of changes to the message caused by operator error and, second, it allowed facsimile signatures to be transmitted, thereby demonstrating the authenticity of a document. However, early facsimile machines also suffered from the significant disadvantage that the image had to be transferred to a conducting plate. Although the following years brought numerous improvements in design, facsimile machines continued to be based upon Bain's contact scanning technique.

It was not until 1902 that the first photo-electric scanning system was developed in Germany by Dr Arthur Korn. The system used the light-sensitive element selenium to convert the different tones of the scanned image into a varying electric current. The image, in the form of a photographic negative, was mounted on a glass cylinder through which light shone. A positive form of the image was printed on sensitized paper at the receiver. This revolutionary system had several significant advantages. To begin with, it was no longer necessary to transfer the image to a sheet of copper or tin foil before transmission. Second, the scanning speed could be increased, because no physical contact was required with the image at the transmitter. Finally, the system was not restricted to black and white—it could also transmit shades of gray. The first commercial use of Dr Korn's system began in 1907 and by 1910, facsimile links had been established between London, Paris and Berlin. In 1922, Dr Korn successfully transmitted a picture from Rome to New York by radio and a commercial radio-based facsimile service was established between London and New York in 1926.

By the 1920s, telegraph-based facsimile services were being used to transmit newspaper pictures around the world. "Phototelegrams" were subsequently used for images ranging from weather maps to machine drawings and fingerprints. Although these services were still available in the 1970s, they were mainly confined to specialist applications.

In the 1960s, relatively cheap facsimile machines started to became available that could operate over the telephone—rather than the telegraph—network. The introduction of the Group 1 facsimile standard in 1968 was a significant step forwards because it enabled inter-working by fax machines from different manufacturers. By the time the Group 3 standard was agreed in 1980, mass production techniques had dramatically reduced the cost of the equipment and facsimile became an important method of communication for both businesses and individuals. However, this period of success was short-lived; by the start of the twenty-first century, facsimile was being rendered obsolete by email.

# 3 A gatecrasher spoils the party

Sunday June 25, 1876 has gone down in history as the date of General Custer's last stand at the Battle of the Little Big Horn. However, this date also proved to be very significant for a young Scotsman by the name of Alexander Graham Bell.

Bell was exhibiting his newly invented telephone at the Centennial Exhibition in Philadelphia, which had been organized to celebrate the 100th anniversary of the American Declaration of Independence. There was no shortage of wonders on display, including the first electric light and a number of important developments relating to the telegraph. As a result, Bell had failed to obtain a stand in the Electrical section and was tucked away in a narrow space between a stairway and a wall in the Education section. After more than 6 weeks on display, his telephone had failed to attract any serious attention.

By a stroke of good fortune, Bell knew one of the commissioners of the Centennial Exhibition. Gardiner G. Hubbard was the father of Bell's girlfriend and had provided Bell with financial support during his years of experimentation. Hubbard somehow managed to persuade a party of the great and the good to pay a visit to Bell's stand—even though it was late in the day, the weather was hot and some of the visitors were understandably reluctant to trek down a corridor and up a flight of stairs to see the new invention.

This reluctance was not shared by Emperor Dom Pedro de Alcantara of Brazil, who had once visited Bell's class of deaf mutes at Boston University and had subsequently helped to organize a school for deaf mutes in Rio de Janeiro. At Bell's invitation, the Emperor picked up the telephone receiver and Bell recited the famous soliloquy from Hamlet, "To be, or not to be ...". "My God, it speaks," exclaimed the Emperor in amazement. Within a few minutes, the telephone had been demonstrated to such influential scientists as Joseph Henry[1] and Sir William Thomson (later to become Lord Kelvin). By the next morning, the telephone had been given pride of place in the exhibition.

We noted in Chapter 1 that Samuel Morse was an unlikely inventor of the telegraph, having spent the first 46 years of his life pursuing a career as a portrait painter. Bell, by contrast, was supremely well equipped to be the inventor of the telephone. Not only was he fascinated by technology—surely an essential requirement for success in this field—but he was also an expert in the science of speech.

Alexander Graham Bell was born in Edinburgh on March 3, 1847. His interest in speech came from other members of his family. His grandfather, also named Alexander Bell, had published writings that included *The Practical Elocutionist* and *Stammering and Other Impediments of Speech* and had been used by George

A. Wheen, *From Dot-Dash to Dot.com: How Modern Telecommunications Evolved from the Telegraph to the Internet*, Springer Praxis Books, DOI 10.1007/978-1-4419-6760-2_4,
© Springer Science+Business Media, LLC 2011

Bernard Shaw as a model for Professor Henry Higgins in his play, *Pygmalion*.[2] His uncle, David, was a teacher of speech in Dublin and his father, Melville, developed "Visible Speech", the first international phonetic alphabet. In Visible Speech, a series of diagrams were used to represent specific actions of the lips and tongue. As a boy, Alexander had helped his father to demonstrate the system. While Alexander was out of the room, someone would be asked to invent a strange or unusual sound; Alexander would then be able to accurately recreate that sound after being invited back into the room and shown the corresponding Visual Speech diagrams. The method could be used to help people learning a foreign language or those wishing to speak their own language more correctly. However, it also had applications in teaching speech to the deaf because it helped people to learn how to make a specific sound even if they could not hear it.

Alexander's mother, Eliza Grace Simmonds, was nearly 10 years older than her husband. She had started to lose her hearing at the age of 12 and her efforts to hear would leave a lasting impression on her son. Most people spoke to Mrs Bell through her ear tube, but Alexander discovered that he could communicate with her by pressing his lips against her forehead so that the bones resonated in response to his voice.

While on a visit to London, Alexander was fascinated by a demonstration of Sir Charles Wheatstone's "speaking machine". Working with his older brother, the pair created an apparatus representing all the main organs involved in human speech. The tongue was made of several small coated paddles, while the larynx came from a dissected sheep and a set of bellows provided the lungs. This apparatus apparently produced surprisingly human-like sounds and a neighbor who heard it say "mama" was firmly convinced that there was a baby in the house.

Bell enrolled at the University of London to study Anatomy and Physiology, but his time at university was cut short. Tuberculosis was rampant in Edinburgh, where his family were living, and both of his brothers died from the disease within the space of 4 months. In an effort to escape from this threat, the surviving members of the family moved to Canada in 1870. Bell himself spent some time recovering from tuberculosis when they arrived at Brantford, Ontario.

In 1871, Bell was invited to visit Sarah Fuller's School for the Deaf in Boston to train some of the teachers in the Visible Speech technique that his father had invented. In 1873, he was appointed Professor of Vocal Physiology and Elocution at the University of Boston, where he taught for the next 4 years. He also opened a "School of Vocal Physiology", and the success of this venture left him with little time for his experiments. However, by a happy coincidence, two of his pupils gave him the motivation and the means to resume his research.

The first was a 15-year-old girl called Mabel Hubbard. She had lost her hearing after an attack of scarlet fever as a baby and, as a result, her speech development had also been seriously affected. Bell was captivated by her and 4 years later, she became his wife. Her father, Boston attorney Gardiner Hubbard, became one of Bell's financial backers and would later provide Bell with his big opportunity at the Centennial Exhibition. Hubbard may not have regarded the impoverished inventor as an ideal son-in-law, but he was determined that his daughter would continue to live in the style to which she had become accustomed.

The second pupil was a 5-year-old boy called Georgie Sanders, to whom Bell had agreed to give a series of private lessons. The Sanders family lived 16 miles outside Boston, so it had been agreed that Bell would live with them. To his delight, they allowed him to use the cellar of their house as a workshop for his experiments. Georgie's father, Thomas, owned a small business that produced soles for shoe manufacturers and he soon became Bell's primary financial backer. In the years immediately following the award of Bell's patent, Thomas Sanders was nearly ruined by the financial demands of the new telephone business.

Bell's work in teaching the deaf to speak stimulated his interest in devices that would enable speech to be visualized. The idea was that such devices could provide deaf people with the feedback that would normally be provided by hearing. If they could see the patterns of their speech and compare them with a visual template representing the required sound, then the deaf person would be able to adjust their pronunciation until the sound matched the template. In 1874, Bell saw two remarkable machines—the manometric flame and the phonautograph—that provided a visible representation of sound. In the first device, a gas flame vibrated in response to the sound of the human voice and the waveform could be viewed as a continuous wavering band of light in a revolving mirror. The second device consisted of a horn terminated by a diaphragm. The diaphragm was linked to a bristle that rested lightly on a moving sheet of paper covered in lamp black. When speech caused the diaphragm to vibrate, the bristle traced out the sound waveform on the sheet of paper.

Bell discussed these devices with a Boston friend, Dr Clarence J. Blake, who suggested that Bell could make a variant of the phonautograph by using a real human ear. Dr Blake was a surgeon and he supplied Bell with an ear removed from a dead man, complete with the ear drum and the associated bones. Bell took this rather grisly offering and while on holiday at Brantford in the summer of 1874, constructed an "ear phonautograph". He arranged the ear so that one end of a piece of straw rested against the ear drum while the other end rested on a piece of smoked glass. When Bell spoke into the ear while moving the smoked glass, he found that the speech waveform was traced out by the piece of straw.

These experiments helped Bell to grasp the basic principles of telephony. Speech is transmitted through the air by fluctuations in pressure. When a sound wave encounters a membrane such as the ear drum, it will cause that membrane to vibrate. If that vibration can be converted into fluctuations in an electric current, then the speech signal can be transmitted down a wire. Taking things one stage further, if the fluctuating electric current can be used to cause vibration in a second membrane located at the receiver, then an audible speech signal will be recreated.

However, a correct understanding of the theory is not sufficient to guarantee success and Bell concluded in November 1873 that his practical skills were not up to the task:

"It became evident to me, that with my own rude workmanship, and with the limited time and means at my disposal, I could not hope to construct any better models. I therefore from this time devoted less time to practical experiment than to the theoretical development of the details of the invention."[3]

Bell recruited a young electrician called Thomas Watson to supply the practical skills that he lacked.

*   *   *

The invention of the telephone came about as a direct result of Bell's attempts to improve the telegraph. As we saw in the previous chapters, the early telegraph pioneers encountered many problems with cables. Since the cost of installing and maintaining cables was so high, there was a strong commercial incentive to reduce the number of cables required to carry the traffic on busy routes.

A device that could transmit two messages over the same wire had been patented by Joseph Stearns in 1872, and Edison had patented a quadruplex telegraph in 1874. It was clear to Bell (and a number of other inventors) that there was money to be made from a device that could further increase the number of simultaneous transmissions over a telegraph wire. Mabel Hubbard's father, Gardiner Hubbard, saw this as an opportunity to break the monopoly of the Western Union Telegraph Company. In 1873, he and Georgie Sanders's father, Thomas Sanders, formed the Bell Patent Association to support Bell's search for an improved form of telegraph.

Bell's initial approach was based upon the work of Hermann von Helmholtz in Germany. Helmholtz had built a machine in which a set of tuning forks and resonant chambers had been used to recreate the vowel sounds that occur in speech. Bell could not read Helmholtz's papers in the original German, so he had to rely on translations. This led to a serious (but fortunate) misunderstanding. Helmholtz's aim had been to generate vowel sounds artificially, but Bell gained the impression that Helmholtz had been *transmitting* vowel sounds electrically over a wire. This led Bell to wonder whether the machine might also be able to transmit consonants, thereby opening the way to a machine that could transmit speech. Bell would later claim that he might never have begun his experiments in electricity if he had been able to read German!

However, Bell was not being funded to invent the telephone—his objective was to improve the telegraph—and Helmholtz's use of electromagnets to keep his tuning forks in oscillation suggested the basis for a "harmonic telegraph" in which a series of telegraph messages could be conveyed simultaneously along the same piece of wire. Bell imagined a set of tuning forks at the transmitter (each tuned to a different frequency) with an identical set of tuning forks at the receiver. If the vibration of each tuning fork at the transmitter could be converted into an electrical signal, then these individual signals could be added together and transmitted down the line. At the receiver, the combined electrical signal could be passed through an electromagnet to generate a magnetic field that would drive a second set of tuning forks. This magnetic field would contain all the frequency components generated by the tuning forks at the transmitter; however, a tuning fork can only vibrate at one frequency, so each tuning fork at the receiver would only respond to the component of the signal generated by the corresponding tuning fork at the transmitter.

This is rather like the action of a radio receiver. Although the radio's aerial is picking up radio signals from many different transmitters, the receiver can be tuned to select just one signal and filter out the rest. In the same way, each tuning fork at the receiver is copying the behavior of the corresponding tuning fork in the transmitter and is completely unaffected by the signals generated by tuning forks operating at different frequencies.

A diagram of one of Bell's original experiments with tuning forks is provided in Figure 6. For simplicity, only one pair of tuning forks is shown.

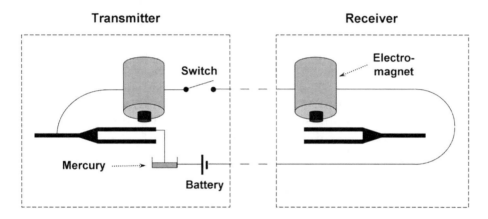

**Figure 6.** Experiment with tuning forks.

One arm of the tuning fork at the transmitter is touching the surface of a bath of mercury, causing it to make and break the electrical circuit as the tuning fork vibrates. This produces an intermittent electric current, which can be used to drive the electromagnet at the receiver. If the tuning fork at the receiver is tuned to the same frequency as the tuning fork at the transmitter, then the fluctuating magnetic field generated by this electromagnet will cause the receiver tuning fork to resonate. Including a switch (or telegraph key) in the circuit allows the vibration of the receiving tuning fork to be turned on and off, thereby enabling messages to be sent down the line in Morse Code.

Bell subsequently replaced each tuning fork with a spring steel "reed"[4] vibrating over the poles of an electromagnet. The pitch of the reed could be tuned by adjusting the length that was free to vibrate. This version of the experiment is shown in Figure 7.

In the apparatus shown in Figure 6, a bath of mercury had been used to make and break an electrical circuit at the transmitter. In the apparatus shown in

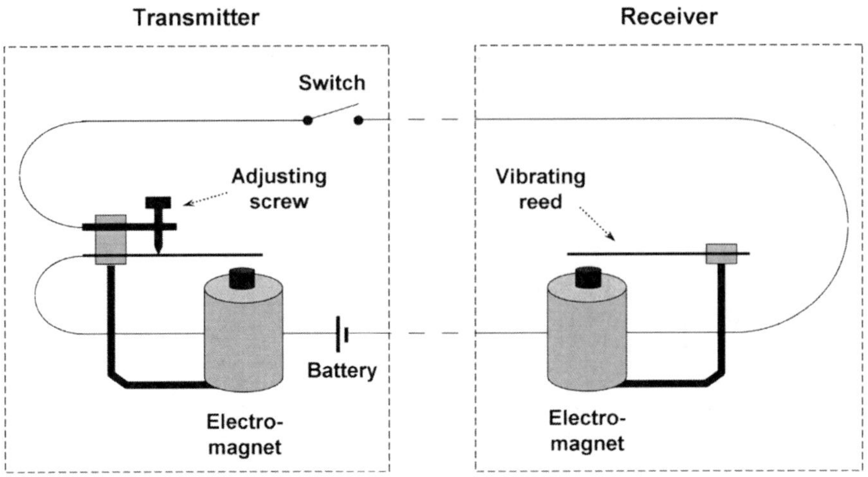

**Figure 7.** Experiment with tuned reed.

Figure 7, an adjustable screw was used to make intermittent electrical contact with the reed as it vibrated (rather like an electric buzzer). In both cases, the electric current was switched on and off at a rate reflecting the frequency of vibration. The point to notice here is that the current has only two possible states: on or off.[5] If a harmonic telegraph is to be made to work, then some way of combining these currents at the transmitter—and then separating them again at the receiver—needs to be found.

In his famous patent of 1876, Bell goes to some lengths to explain the problems of trying to combine a number of "pulsatory" currents of this type on a single wire. His argument is surprisingly simple. If two telegraph signals are carried on a single wire, then current will flow for a higher proportion of the time than would be the case if just one of the signals was being carried, because current will flow when either of the two switches is closed. If three telegraph signals are combined together, then current will flow for an even higher proportion of the time, because there is an increased probability that at least one switch will be closed at any given moment. Using an argument of this type, it is possible to show that a small number of telegraph signals can be combined together to create a current that flows almost continually. Clearly, no useful information will be recovered at the receiver if this occurs, because it is the breaks in the current that define the dots and the dashes.

Bell soon recognized that the key to the harmonic telegraph was to work with "undulatory" current. Unlike "pulsatory" currents, which are either on or off, undulatory currents can take any value within a given range.[6] Under the right circumstances, undulatory currents can be summed together without any information being lost. Figure 8 shows two undulatory signals—A and B—being added together to produce a third signal (A+B). With suitable filtering at the receiver, it is possible to separate the combined signal back into its two component parts.

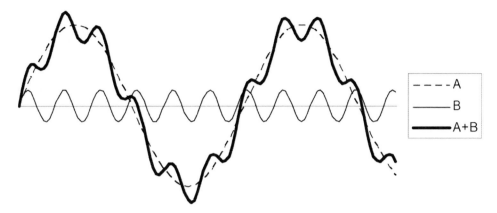

**Figure 8.** Summing undulatory currents.

When examining Bell's sequence of experiments between 1873 and 1876, it is easy to be confused as to his true aim. This is illustrated by his response to the work of Helmholtz. On the one hand, he saw the set of tuning forks kept in oscillation by electromagnets as providing the basis for a harmonic telegraph. On the other hand, he was wondering whether a device that could (as he mistakenly thought) transmit vowel sounds could also be used to transmit other speech components. Although the funding provided by Gardiner Hubbard and Thomas Sanders was intended specifically to develop a harmonic telegraph, Bell became increasingly obsessed with the idea of developing a telephone.

Hubbard and Sanders could see the commercial potential of a harmonic telegraph, but they believed that a telephone—even if it could be made to work—would never be more than a laboratory curiosity. At one point, they even threatened to cut off funding unless Bell confined his researches to the harmonic telegraph. Bell's poverty ensured that he complied, but the problems of the telephone were never far from his mind. Finally, on June 2, 1875, an incident occurred that caused him to abandon his work on the harmonic telegraph and focus all his energies on the development of the telephone.

Bell and Watson were experimenting with a circuit containing a number of tuned reeds. The idea was that the behavior of a transmitting reed would be copied by any reed tuned to the same frequency but would be ignored by reeds tuned to different frequencies. The transmitter reeds were set up to vibrate like electric buzzers. However, as Watson would later describe, the experiment did not produce the expected results:

> "About the middle of the afternoon, we were retuning the receiver reeds, Bell in one room pressing the reed against his ear one by one as I sent him the intermittent current of the transmitters, from the other room. One of my transmitter reeds stopped vibrating, I plucked it with my fingers to start it going. The contact point was evidently screwed too hard against the reed and I began to readjust the screw while

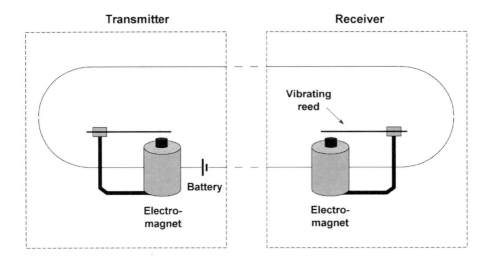

**Figure 9.**   Simplified experimental configuration.

continuing to pluck the reed when I was startled by a loud shout from
Bell and out he rushed in great excitement to see what I was doing."[7]

The adjusting screw had been done up too tightly, so the transmitter reed was
never actually breaking the electrical circuit. If we ignore the adjusting screw and
the other transmitters and receivers—none of which had any bearing on this
result—the experimental configuration used by Bell and Watson can be reduced
to the much simpler circuit shown in Figure 9.

This diagram shows two identical electromagnets connected together in a
circuit. A reed is clamped horizontally above each electromagnet in such a way
that one end is free to move up and down in the magnetic field created by the
electromagnet. When a battery is included in the circuit, both electromagnets
are activated and the reeds are drawn downwards by magnetic attraction.

Whenever a piece of steel is attracted by a magnet, that steel becomes a
magnet in its own right. This means that each reed must now be regarded as a
magnet.[8] If the reed at the transmitter is plucked so that it starts to vibrate, we
have the situation in which a magnet is moving close to a coil of wire and
Faraday demonstrated in 1831 that this causes a current to flow in the coil. As the
reed moves closer to the electromagnet, the induced current will flow in one
direction; when the reed moves away from the electromagnet, the current will
flow in the opposite direction.[9] This electric current provides a direct
representation of the vibration of the reed. Since variations in the current will
cause corresponding variations in the magnetic field of the receiver electromagnet, the receiver reed will follow the vibrations of the transmitter reed. By
pressing his ear against the receiver reed, Bell was able to hear the twanging of
the transmitter reed as it was plucked.

This was one of those happy accidents that sometimes lead to major scientific breakthroughs. The important point was not that the reed failed to behave as expected—scientific experiments often go wrong—but that Bell was able to recognize the true significance of what had happened. Bell had known for some time that an undulating electric current could be generated using an electro-magnet, but he had assumed that the current would be too weak to transmit speech down a wire. This experiment suggested that his assumption was wrong. Furthermore, Bell found that he obtained the same result when other reeds (tuned to different frequencies) were plucked. This suggested that a single reed could transmit all the frequencies required to reproduce a complex speech waveform—the large number of separately tuned devices used by the Helmholtz machine were not necessary.[10]

This experiment convinced Bell that the telephone could be made to work and he immediately asked Watson to construct a prototype. A version of the illustrative design that he subsequently included in his patent application is reproduced in Figure 10.

In this design, the two electromagnets have been placed on their sides so that the reeds are facing outwards. At the transmitter side, the reed is attached to the middle of a stretched membrane, which is based on the design of the human ear drum. A cone is used to direct the speaker's voice towards the membrane, thereby causing it to vibrate. This, in turn, causes the attached reed to vibrate, generating an undulatory current corresponding to the speech waveform.

This undulatory current flows along the wire to the receiver, where it causes fluctuations in the magnetic field generated by the right-hand electromagnet. These fluctuations cause the reed at the receiver to vibrate and these vibrations are coupled to the middle of a second membrane. Since the reed in the transmitter is effectively coupled to the reed in the receiver, this means that there is also a coupling between the two membranes. Furthermore, since the movement of the first membrane is driven by the speech waveform that it is detecting, the movement of the second membrane will reproduce the same waveform: sound detected at the transmitter will be reproduced at the receiver.

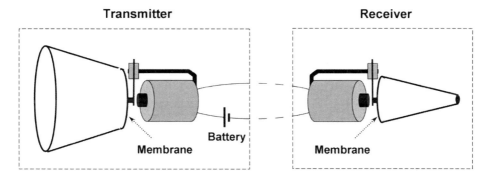

**Figure 10.** Telephone prototype design.

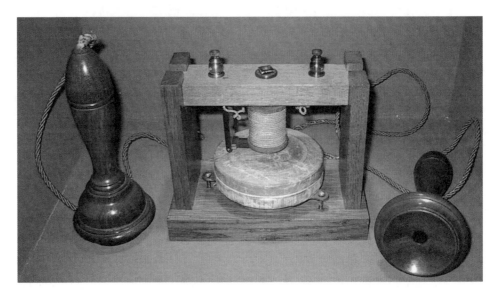

Reproduction of gallows telephone.

The telephone that Watson produced has since become known as the "gallows telephone" because of a slight resemblance to a hangman's scaffold when it is turned on end. This device did not transmit intelligible speech, but the results were sufficiently promising to convince Bell that they were on the right track.[11] During the next 9 months, he filed an application for a patent and continued with his experiments. However, his attempts to create a working telephone were not successful.

In the spring of 1876, Bell designed a new type of transmitter that generated an undulating waveform in a different way. A wire attached to a membrane was inserted into battery acid contained in a metal cup. As the membrane vibrated, the wire would be moved up and down in the liquid, causing the resistance between the wire and the cup to fluctuate. This fluctuating resistance would cause an undulatory current to flow in the circuit.

Bell's telephone patent was issued on March 7, 1876. Three days later, the new transmitter design was tested and an extract from Bell's laboratory notebook for that day reads as follows:

*March 10, 1876*
"The improved instrument shown in Fig. 1 was constructed this morning and tried this evening. P is a brass pipe and W a platinum wire, M the mouth piece and S the armature of the receiving instrument.

Mr Watson was stationed in one room with the Receiving Instrument. He pressed one ear closely against S and closed his other ear with his hand. The Transmitting Instrument was placed in another room and the doors of both rooms were closed.

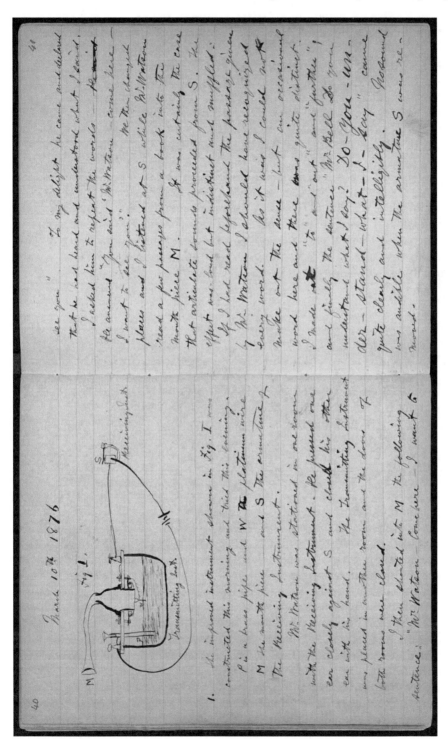

Extract from Bell's laboratory notebook for March 10, 1876.

I then shouted into M the following sentence 'Mr Watson—come here—I want to see you'.[12] To my delight he came and declared that he had heard and understood what I said.

I asked him to repeat the words. He answered 'You said "Mr Watson—come here—I want to see you"'. We then changed places and I listened at S while Mr Watson read a few passages from a book into the mouth piece M. It was certainly the case that articulate sounds proceeded from S. The effect was loud but indistinct and muffled.

If I had read beforehand the passage given by Mr Watson I should have recognised every word. As it was I could not make out the sense but an occasional word here and there was quite distinct. I made out 'to' and 'out' and 'further', and finally the sentence 'Mr Bell do you understand what I say? **Do—you—under—stand—what—I say**' came quite clearly and intelligibly. No sound was audible when the armature S was removed."[13]

Three months later, Bell's telephone caused a sensation at the Centennial Exhibition in Philadelphia, and the Bell Telephone Company was established in the following year. However, Bell soon lost interest in the further development of his invention. When the Bell Telephone Company offered him a salary of $10,000 a year to remain its chief inventor, he refused the offer on the grounds that he could not invent to order. Rich, famous and barely into his thirties, Bell retired to his country house in Baddeck and turned his fertile mind to inventions in other fields. These included:

Alexander Graham Bell.

- a vacuum jacket that administered artificial respiration when placed around the chest (in response to the death of his baby son from breathing difficulties);
- the photophone, which transmitted speech using a beam of light instead of a wire;
- a metal detector that could detect bullets and other metal objects in the body (in response to the assassination of President James Garfield in 1881);[14]

- experiments in sheep breeding to produce ewes that would always bear twins.[15]

During his life, Bell was granted 18 patents in his own name and shared another 12 with collaborators. His last patent, granted when he was 75, was for a new form of hydrofoil.

\* \* \*

Shortly after the invention of the telephone, Bell had written to his father "The day is coming when telegraph wires will be laid on to houses just like water or gas—and friends will converse with each other without leaving home". Bell was considerably more far-sighted than most of his contemporaries—at the time, the telephone was widely dismissed as little more than a toy. Many regarded it as an interesting but specialized variant of the telegraph—rather like Bain's facsimile telegraph. Bell himself described the telephone as a telegraph machine that did not require experts to operate it.

Although the telephone was, in one sense, a natural development of the telegraph, that is rather like saying that the automobile was a natural development of the horse and cart. In both cases, the development may have been natural from a technology point of view but it opened up a whole new range of possibilities that had not existed before. This was illustrated by the rapid take-up of the telephone. By the end of June 1877, there were 230 telephones in use. This figure grew by over 1,000 during the following 2 months and by 1880, there were 30,000 telephones in use around the world. By 1886, the number had risen to over 250,000.

The telephone eventually became one of the most valuable patents ever issued. Not surprisingly, it triggered a flood of lawsuits. As in the case of the telegraph, a number of people had been conducting experiments in related fields at about the time that Bell filed his patent. Three years after the patent was granted, 125 competing companies had been established in open defiance of the patent.

The first attack on the Bell patent—and probably the most dangerous—came from the Western Union Telegraph Company. By 1866, Western Union controlled 90% of the telegraph traffic in the United States, having swallowed up dozens of smaller telegraph companies. Since then, the company had grown rich and powerful on monopoly profits. Bell described the company in a letter to his parents as "probably the largest corporation that ever existed".

In 1877, Bell had offered his telephone patent to Western Union for $100,000, but they had scornfully rejected the offer.[16] However, they soon realized their mistake and their senior electrical expert, Frank L. Pope, was given the task of finding a way around the Bell patent. After 6 months of searching libraries and patent offices across Europe and the United States, Pope was forced to admit defeat. He recommended that Western Union should try to purchase the Bell patents. This recommendation was not well received by the senior management at Western Union. How could an impoverished start-up resist the might of a corporate giant? Before long, Western Union had launched a telephone service in direct competition to Bell, having first recruited a number of prominent inventors to help them circumvent the Bell patent. Although Western Union's position was legally dubious, they felt secure in the well established principle that "might is right".

One of the prominent inventors recruited by Western Union was Thomas Edison, who later went on to find fame and fortune for a string of other inventions; his contribution to the Western Union cause is described in Appendix C. Another was Elisha Gray, who, for some years, had been one of Bell's fiercest competitors. Both Gray and Bell had started working on the harmonic telegraph, but Gray had continued with this work after Bell had redirected his efforts towards the telephone. Each suspected the other of stealing ideas. Gray had submitted a patent application relating to the harmonic telegraph on February 23, 1875, 2 days before Bell filed a similar application. In the case of the telephone, the roles were reversed: Gray filed his caveat application[17] a mere 2 hours after Bell had filed his patent!

Careful examination of Gray's caveat shows that he understood the principles of the telephone and—given a little more time—would almost certainly have been able to produce a working model. Furthermore, Gray's caveat contained a description of a variable resistance transmitter that was remarkably similar to the one used by Bell when he spoke the first words to be received over a telephone. It was later claimed that the part of Bell's patent that deals with the variable resistance transmitter was added after it had been filed. Gray believed that Bell had updated his patent application after illegally being shown a copy of Gray's caveat by friends in the Patent Office.

Not surprisingly, legal hostilities soon broke out between Western Union and the Bell Telephone Company, and the case finally came to court in the autumn of 1878. Western Union had expected their legal firepower to win the day, so it must have come as a nasty shock when, 1 year later, they were forced to seek a settlement. This occurred because their senior lawyer, George Gifford, became convinced that the Bell patent was valid. As one of the leading patent attorneys of his day, Gifford's views could not be easily dismissed.

After months of acrimonious negotiation, an agreement was finally reached. Under the terms of this agreement, Western Union admitted that Bell was the original inventor of the telephone and that his patents were valid. In return, the Bell Telephone Company agreed to buy Western Union's telephone system and to pay them a royalty of 20% on all telephone rentals. Crucially, it was also agreed that Western Union would no longer attempt to enter the telephone business while the Bell Telephone Company would stay out of the telegraph business.

This agreement remained in force for the remaining 17 years of the Bell patent and was a staggering victory for the fledgling Bell Company. It added 56,000 telephones to the Bell system and converted a very dangerous competitor into a powerful ally. Stock in the Bell Telephone Company soared to $1,000 per share. However, Bell's legal problems were far from over.

The next significant challenge to Bell's patent came from Professor Amos E. Dolbear, who claimed that Bell's invention was simply an improvement upon a device constructed in Germany in 1860 by Philip Reis. Reis had actually been trying to develop a telephone, but he found that his device was only capable of transmitting simple tones. The transmitter consisted of a lever resting on a piece of platinum in the center of a membrane. A musical note would cause the membrane to vibrate, thereby making and breaking the electrical contact between the lever and the piece of platinum. This caused an electrical current to be switched on and off with a frequency corresponding to the frequency of the tone. Passing this current through an electromagnet could be used to set up vibration in an adjacent piece of metal, thereby reproducing the tone.

The Reis machine was almost certainly known to both Bell and Gray, and would have helped to stimulate their thinking. However, the machine was based upon a fundamentally flawed principle—it switched the electric current on and off in response to sounds, rather than attempting to create an accurate representation of the sound waveform. In the terminology of Bell's patent, it produced a pulsatory rather than an undulatory current. It could transmit pure tones relatively well, because the ear can detect the original tone in a pulsatory current (square wave) of the same frequency. However, it proved to be hopeless at transmitting speech, in which much more complex waveforms have to be handled. As the judge stated when delivering his ruling against Professor Dolbear's claim:

> "A century of Reis would never have produced a telephone by mere improvement of construction. It was left to Bell to discover that the failure was due not to workmanship but to the principle which was adopted."[18]

In total, the Bell Telephone Company successfully defended their patent against more than 600 legal challenges over a period of 11 years. Five of these challenges went all the way to the Supreme Court. One of their strongest arguments was the historical fact that Bell's telephone had been seen at the Centennial Exhibition by some of the most eminent electrical scientists of the day and they had all declared it to be something completely new.

In view of this, it is perhaps surprising that Bell's claim to have invented the telephone was rejected by the US Congress more than a century after the main legal arguments had been settled. On September 25, 2001, Congress passed a resolution that officially recognized a Florentine immigrant to the United States, Antonio Meucci, as the inventor of the telephone. On December 28, 1871, Meucci had filed a caveat for his "teletrofono" because he could not afford the $250 required for a full patent application. By 1874, he could not even afford the $10 needed to renew the caveat. If it had not been for Meucci's extreme poverty, it appears that Alexander Graham Bell would not have been granted his patent 2 years later.

Meucci later joined the ranks of those initiating legal action against Bell, claiming that Bell had shared a laboratory with him and was aware of his work.

However, Meucci died before the case could be brought to a successful conclusion.

So, who really did invent the telephone? As we saw in the case of the telegraph, the answers to such questions are rarely simple or clear-cut. There are two main reasons for this:

Antonio Meucci.

1. *Inventors build upon the work of other inventors.* For example, Samuel Morse built upon the work of another American, Joseph Henry, who had demonstrated in 1830 that it was possible to send an electric current over a distance of 1 mile and cause a bell to ring. However, the electric bell would not have been possible without the discovery of the electromagnet in 1825 by the British inventor, William Sturgeon. This, in turn, depended upon the discoveries of scientists such as Michael Faraday, who investigated the linkage between electricity and magnetism.

2. *A number of inventors are likely to be working on the same problem at the same time.* Within any given field of endeavor at any given moment in time, there is a level of understanding among the foremost practitioners that represents the "state of the art". We should not, therefore, be surprised if a number of inventors, striving to advance the frontiers of knowledge, come up with the same idea at about the same time. By 1876, the telephone was an idea whose time had come. Bell's patent application was submitted only 2 hours before Elisha Gray tried to patent his microphone system. Others, such as Thomas Edison, were also active in the field.

In some senses, men like Morse and Bell had the good fortune to be able to place the final link in a chain that had been forming over a number of years. They were certainly not the first to envisage the basic concepts that lay behind their inventions—Charles Morrison had proposed a telegraph system 75 years before Morse turned his attention to the subject and Charles Bourseul published the idea of the electrical transmission of sound 22 years before Bell's patent. They were not even the first to provide a physical realization of that concept— Cooke and Wheatstone opened the world's first public telegraph service between Paddington and West Drayton 5 years before Morse's experimental system between Washington and Baltimore, and Meucci gave a public demonstration of a telephone 16 years before Bell's appearance at the Centennial Exposition.

However, it is undeniably true that both Morse and Bell contributed crucial refinements that converted their inventions from laboratory curiosities into practical technologies that could be deployed to solve real-world problems. The widespread adoption of the telegraph—which led directly to the development of the telephone—has brought untold benefits to our world. For that, men such as Morse and Bell deserve our grateful thanks.

# 4 Early telephone networks

In May 1877, an advertisement was produced for Bell's newly invented telephone. It modestly claimed that the telephone was superior to the telegraph for three reasons:

> "(1) No skilled operator is required, but direct communication may be had by speech without the intervention of a third person.
> (2) The communication is much more rapid, the average number of words transmitted in a minute by the Morse sounder being from fifteen to twenty, by telephone from one to two hundred.
> (3) No expense is required, either for its operation or repair. It needs no battery and has no complicated machinery. It is unsurpassed for economy and simplicity."

However, a cynic might have pointed out that the telegraph network enabled messages to be sent around the world, while no telephone network existed at all! Early telephone lines simply provided a link between two locations—no dial was needed because there was only one person you could call![1] This chapter examines how the telephone developed from a few short-distance, point-to-point links into a global network that would challenge the dominance of the telegraph.

Let's start by considering some basics. One approach to building a network would be to provide a direct connection from every subscriber to every other subscriber on the network. Although this is possible in theory, it soon becomes unmanageable in practice. For example, to add a new subscriber to a system already serving 10,000 subscribers would require 10,000 new connections—one to each of the existing subscribers. The cost of doing this would be far greater than the revenue that the new subscriber would ever be likely to generate. Furthermore, the need to terminate 10,000 separate telephone wires at each subscriber's house would be hopelessly inconvenient.

Clearly, it is impractical to provide *dedicated* resources to support every possible telephone call that might be made. What is needed is a *shared* pool of network resources that are only allocated to an individual subscriber when they are needed. Since most subscribers only spend a small proportion of their time on the telephone (apart from teenagers, of course), there is no need to build a network with so much capacity that every subscriber can use the telephone all the time.

It is rather like the national road network. Nobody owns their own private road network—it is a shared resource that everyone uses. The only part of the road network that belongs to an individual user is the driveway that connects the

A. Wheen, *From Dot-Dash to Dot.com: How Modern Telecommunications Evolved from the Telegraph to the Internet*, Springer Praxis Books, DOI 10.1007/978-1-4419-6760-2_5, © Springer Science+Business Media, LLC 2011

road to their garage. In the case of the telephone network, the only part of the system that is dedicated to an individual subscriber is the telephone line connecting that subscriber to the local exchange,[2] plus a small amount of terminating equipment in the local exchange. The majority of equipment in a telephone network is shared by the users of the network.

Pursuing the road analogy a bit further, we can be certain that if every car and lorry in Britain were to attempt to travel from London to Birmingham at 9 a.m. on a Friday morning, the result would be chaos. The M1 is a very fine road, but it was never designed to carry this level of traffic. Fortunately, the road planners can safely ignore this problem (and others like it) because it is never likely to happen. The M1 has—at least in theory—been designed to carry the volume of traffic that normally needs to use it. The planners also know that the volume of traffic traveling from London to Birmingham at any given time is likely to be far higher that the volume of traffic traveling between Ballachulish and Kinlochleven in the Scottish highlands, so the M1 has a much higher traffic capacity than the B863. It is the same with the telephone network. The trunk lines between London and Birmingham are capable of carrying thousands of simultaneous phone calls, while the connection to a small village in a remote part of the country might be restricted to fewer than 100.

Of course, all this planning can break down when something unusual occurs. In the road network, large rock concerts and major sporting events can generate massive traffic congestion in their immediate vicinity, and unexpected call patterns can produce similar problems on the telephone network. Consider, for example, what happens when a popular TV talent show invites its viewers to vote by calling a particular telephone number and thousands of people then try to call that number simultaneously. If the telephone network operator has been warned in advance, then steps can be taken to handle the surge of calls that will occur. If, however, no warning has been given, then the result is likely to be massive network congestion.

Although the early telephone pioneers did not have to deal with the problems caused by TV talent shows, they did need to minimize the amount of network equipment that was dedicated to an individual subscriber in order to keep down costs. The amount of shared equipment in the network had to be sufficient to carry all *likely* call patterns, but they could not afford to build additional capacity that would only be used under extreme conditions. Fortunately, the telegraph network was already a large and well established international network by the time that telephone networks started to appear, so many of the lessons learned on the telegraph could simply be re-applied. It is therefore worth reviewing how these problems were solved in the case of the telegraph before considering how the telephone pioneers built their networks.

Local telegraph offices were often connected together in lines. Each office could see all the traffic passing along the line, but only noted down the messages that were addressed to their specific area. Each local telegraph office could send messages directly to any other telegraph office connected to the same telegraph line. This is illustrated by Row A in Figure 11.

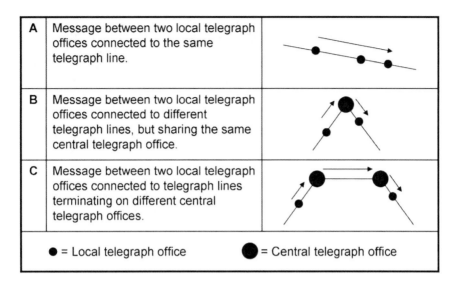

| A | Message between two local telegraph offices connected to the same telegraph line. | |
| B | Message between two local telegraph offices connected to different telegraph lines, but sharing the same central telegraph office. | |
| C | Message between two local telegraph offices connected to telegraph lines terminating on different central telegraph offices. | |
| ● = Local telegraph office | | ⬤ = Central telegraph office |

**Figure 11.** Message routes between telegraph offices.

Telegraph lines were brought together—like the spokes of a wheel—at the central telegraph office in a large town. A message to a telegraph office on a different telegraph line had to be sent via the central telegraph office, as illustrated by Row B in Figure 11. The sending office would transmit the message to the central telegraph office and the central telegraph office would then forward the message to the receiving telegraph office.

Central telegraph offices were staffed with large numbers of people whose job was to receive messages, work out where they were going and then send them on their way. In the simplest case, an operator at the central telegraph office would have to re-key each incoming message. Later on, the introduction of punched paper tape technology meant that a message could be forwarded by simply transferring the paper tape from one machine to another.

If the sending and receiving telegraph offices were not connected to the same central telegraph office (as illustrated by Row C in Figure 11), then the message would have to be transmitted in three stages. Of course, if a direct connection did not exist between the two central telegraph offices, then transmitting the message from end to end could require four or more separate stages.

In the very early days of the telephone, an arrangement similar to the one shown in Row A of the figure was tried, with limited success. Small groups of telephone subscribers were linked together so that they could talk to each other using a shared line. However, this meant that all the telephones on the system rang whenever someone was trying to make a call. To avoid the need to answer calls intended for someone else, each household was allocated a different ringing pattern and the correct ringing pattern had to be used when making a call. The use of a shared line also meant that the system could only support one call at a time. Since anyone could listen in to that call, eavesdropping was a common problem.[3]

The shared line had worked in the case of the telegraph because the telegraph was only used by trained operators. However, as Bell's advertisement made clear, one of the key advantages of the telephone was that it could be used by *anyone*. Shared line systems were soon abandoned in favor of an architecture based upon another piece of telegraph technology—the switchboard.

The telephone switchboard was the forerunner of the modern telephone exchange. A switchboard would be established at a central location in a town and each telephone subscriber in the area would be linked directly to it. The first regular telephone exchange was opened in New Haven, Connecticut, on January 28, 1878.[4]

Each telephone line would terminate at a socket on the switchboard. In order to set up a call, the telephone operator would connect two sockets together using an electrical patch chord. (A patch chord is a short piece of cable with a plug at each end.) The procedure for doing this was as follows:

1. The calling subscriber picked up the telephone. This switched on a lamp above the subscriber's socket on the switchboard.
2. The operator connected her headset to the subscriber's socket and asked the subscriber whom they would like to talk to.
3. The operator disconnected her headset from the calling subscriber's socket and connected instead to the socket belonging to the person being called. A ringing signal was sent down the line to cause their telephone bell to ring.
4. If the telephone was answered, the operator would disconnect her headset and use a patch chord to connect the two sockets together, thereby establishing the call.
5. When the lamps above the sockets went out, the operator knew that the call was finished. The patch chord could then be removed.

This procedure enabled local calls to be made between telephones connected to the same switchboard. However, things become a little more complicated for long-distance calls, because more than one switchboard was involved. Since each switchboard required an operator, let's give them some names: Anne, Betty, Carol and Daisy.[5] The situation is illustrated in Figure 12.

The procedure to set up a long-distance call would be as follows:

1. The calling subscriber picked up the telephone. This switched on a lamp above the subscriber's socket on the switchboard.
2. The operator (Anne) plugged her headset into the calling subscriber's socket and asked the subscriber whom they would like to talk to.
3. Anne disconnected her headset from the calling subscriber's socket and connected it instead to one of the lines going to a long-distance exchange. Here, a second operator (Betty) was waiting to receive the call.
4. Anne told Betty the destination number for the call.
5. Betty connected Anne to one of the lines going to another long-distance exchange. Here, a third operator (Carol) was waiting to receive the call.

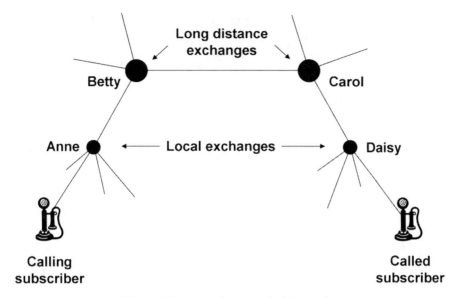

Betty

Carol

Long distance
exchanges

Anne ←——— Local exchanges ——→ Daisy

Calling
subscriber

Called
subscriber

**Figure 12.** Long-distance telephone call.

6. Anne told Carol the destination number for the call.
7. Carol connected Anne to one of the lines going to the required local exchange. Here, a fourth operator (Daisy) was waiting to receive the call.
8. Daisy connected Anne to the telephone line of the called subscriber and the telephone bell was rung.
9. If the telephone was answered, Anne disconnected her headset and used a patch chord to connect the calling subscriber to the long-distance line, thereby establishing the call.
10. When the lamps above the sockets went out, each of the operators knew that the call was finished. The patch chords at each exchange could then be removed.

This may sound rather complicated, but it isn't really—the process followed at each exchange is essentially the same. However, setting up a call in this way is certainly slow, labor-intensive and prone to errors. It might therefore be expected that it would be these considerations that drove the introduction of automation into the telephone exchange. Surprisingly, it was not. The impetus to develop the first automatic telephone exchange came from competitive tensions in the funeral business.

Almon B. Strowger was a Kansas City undertaker. The story goes that one of the operators in the local telephone exchange was the wife of another undertaker. Whenever calls came in for Strowger, she would route them to his competitor—despite Strowger's repeated complaints to the telephone company. This loss of business was obviously a powerful incentive for Strowger to find a way of eliminating the need for a human operator. In 1889, he came up with a mechanism that would allow subscribers to establish their own local telephone

Telephone exchange, 1892.

calls without the involvement of an operator. Two years later, he received a patent for the first automatic telephone exchange.[6] In 1892, he founded the Strowger Automatic Telephone Exchange Company and the first exchange using a Strowger switch was opened in La Porte, Indiana, in the same year.[7] Amazingly, switches based upon Strowger's design were still in use well over 100 years later.

Strowger's telephone exchange used an electromechanical device called a selector. This is a form of rotary stepping switch and is illustrated in Figure 13.

The movement of the selector is driven by an electromagnet. Each pulse of current sent up the line causes the electromagnet to activate briefly and this pushes the selector round by one position.

Let's assume that the calling subscriber (shown on the left of the diagram) wishes to place a call to one of the subscribers shown on the right. Since there are only 10 possible destinations for the telephone call in this simple scenario, a single-digit telephone number is sufficient. When the calling subscriber dials the

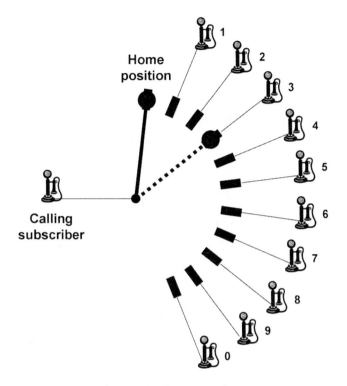

**Figure 13.** Strowger selector.

number, pulses are sent up the telephone line to the selector. The selector is initially in its Home Position, but each pulse causes it to move one position in a clockwise direction. In the situation illustrated, three pulses have been sent up the line, so the selector comes to rest at Subscriber No. 3.

A device of this type is sometimes referred to as a uniselector, because the switch can only move in one plane. However, it is possible to imagine a switch consisting of a number of uniselectors stacked one on top of the other. Such a device is known as a two-motion selector and is illustrated in Figure 14.

The two-motion selector's home position is at the bottom left-hand corner of the switch. When the first digit is dialed, the selector increments upwards by the corresponding number of rows. When the second digit is dialed, the selector

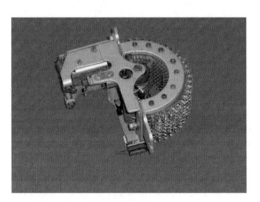

Uniselector.

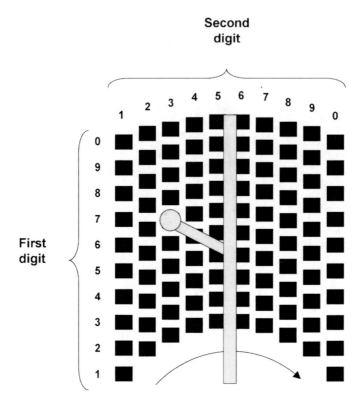

**Figure 14.** Two-motion selector.

rotates clockwise by the corresponding number of places. In the example shown in Figure 14, the subscriber has dialed the number 63.

The uniselector shown in Figure 13 contains the implicit assumption that the subscriber on the left will make calls but never receive them, while the 10 subscribers on the right will only ever receive calls. This, clearly, is not a valid assumption! We need an additional piece of equipment that allows the uniselector to be shared by all the subscribers who might need to use it. One possible arrangement is shown in Figure 15.

When the calling subscriber lifts their handset to make a call, the switch on the left rotates until it connects to that subscriber's line. This means that the uniselector on the right is now assigned to that subscriber for the duration of the call. When the subscriber dials the number 3, the uniselector increments clockwise by three places, as illustrated in the diagram. This connects the subscriber to a further uniselector, which responds to the second number dialed. This, in turn, connects the subscriber to a third uniselector and so on, until the subscriber has been connected to the person that they are calling. In practice, each uniselector might be replaced by a two-motion selector called a group selector. Group selectors step upwards in response to a dialed digit and

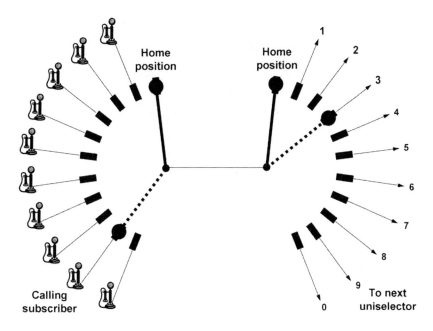

**Figure 15.** Shared uniselector.

then rotate until they find an unused group selector that can receive the next digit.

This is starting to sound like a viable concept for an automatic telephone exchange, but there are still many hurdles to overcome. How, for example, does the exchange handle the situation where the called subscriber's line is engaged or the number dialed is not valid? What happens in Figure 15 if several of the subscribers try to make a call at the same time? How are the uniselectors reset to their home positions at the end of a call? How is the telephone bell made to ring when there is an incoming call? When you start to delve into the details of a telephone exchange, you soon find that this is pretty complicated stuff. Let's call a halt to this discussion and move on, having first concluded that people who design telephone exchanges must be really rather clever.

\* \* \*

In the days when telephone exchanges were based upon switchboards, calls were set up by the calling subscriber talking to the operator in the local exchange and then (if necessary) by the local exchange operator talking to operators in other parts of the network. However, an automatic telephone exchange needs some way to communicate the required telephone number from the calling subscriber to the various parts of the network that will carry the call. Communication of this type between different parts of a telephone network is called "signaling".

Rotary dial. For the benefit of younger readers, who may never have used a rotary dial, this is how they worked. To dial a number (7, for example), the user would put a finger in the hole marked 7 and would then rotate the dial clockwise until their finger encountered the finger stop (the metal flap just to the right of 0). The finger would then be removed and a spring would return the dial to its former position at a steady speed determined by a centrifugal governor. As the dial returned, it would cause a switch to open and close seven times, thereby sending seven pulses up the line. As can be seen from the photograph, dialing a number less than 7 required the dial to be rotated by a smaller amount, so proportionately fewer pulses would be sent up the line. On the dial shown, dialing 0 would result in 10 pulses being sent.[8]

In the original Strowger design, subscribers were required to tap out the number that they wanted on three separate keys (one key for each digit in the telephone number). When the call was finished, the subscriber was required to press a release button to reset the selectors to their home positions. Not surprisingly, this approach proved to be error-prone and the keys were replaced by the rotary dial in 1896.

For almost 100 years, the rotary telephone dial remained the standard way of signaling telephone numbers between the subscriber and the local exchange. Although it was slow, the speed at which the dial rotated was limited by the speed at which the mechanical switches in the local exchange could operate, so there was no point in going any faster. However, when digital technology was introduced into the local exchange, a faster method of signaling was required.

Modern push-button telephones use a signaling system called Dual Tone Multi-Frequency (or DTMF for short). Each digit on the keypad is represented by a pair of tones, as shown in Table 2.

**Table 2.** DTMF signaling tones.

|          | 1,209 Hz | 1,336 Hz | 1,477 Hz |
|----------|----------|----------|----------|
| 697 Hz   | 1        | 2        | 3        |
| 770 Hz   | 4        | 5        | 6        |
| 852 Hz   | 7        | 8        | 9        |
| 941 Hz   | *        | 0        | #        |

You can hear these tones when you pick up the telephone and dial a number. For example, if you press 6 on the keypad, you will hear a 770 Hz tone combined with a 1,477 Hz tone. At the local exchange, a bank of filters detects the presence of these two tones (and the absence of the other five possible tones), indicating that a 6 has been dialed.

The first tone dialers were installed in Baltimore in 1941, but they were reserved for telephone operators, being far too expensive for use by the general public. By the early 1960s, the falling cost of electronics made it possible to offer this form of dialing to subscribers, although it was many years before push-button telephones became commonplace in UK homes.

*   *   *

By the end of the nineteenth century, the invention of the rotary dial and the Strowger automatic telephone exchange meant that it was possible to set up and clear down local telephone calls automatically.[9] However, long-distance calls still had to be set up by dialing 0 and talking to the operator—a situation that persisted in the United Kingdom until 1958, when subscriber trunk dialing (STD) was introduced.[10] Setting up long-distance calls automatically requires signaling not only between the calling subscriber and the local exchange, but also between the local exchange and other exchanges that will be involved in handling the call. Automation of long-distance calls requires some form of inter-exchange signaling.

The links between telephone exchanges are referred to as "trunks". In the early days of the telephone, each trunk would only be able to carry a single phone call, so a large number of cables could be required between busy exchanges. Trunks are shared resources that are available to anyone making a long-distance call, so it is necessary for the telephone exchanges at each end to know whether the trunk is currently busy (i.e. carrying a phone call) or idle. In the early days of subscriber trunk dialing, this was typically done using tones.

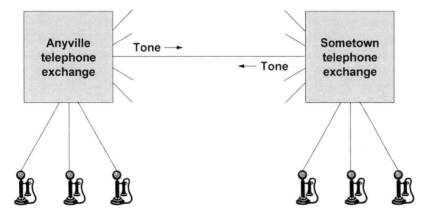

**Figure 16.** Trunk signaling.

Figure 16 illustrates a trunk linking the telephone exchanges at Anyville and Sometown. When the trunk is idle, each exchange indicates this by sending a tone down the line (the trunk is not carrying any telephone conversations, so it might as well carry a signaling tone instead). Now, let's suppose that a subscriber in Anyville wishes to call a friend in Sometown. When the number has been dialed, the Anyville exchange will know that it needs a trunk to carry the call to the exchange at Sometown. It searches until it finds a trunk that is free and then seizes it by turning off the tone. This tells the Sometown exchange that the Anyville exchange is about to use the trunk to send it a telephone call, so the Anyville exchange turns off the tone that it is sending in the other direction. At this point, the Anyville exchange needs to transfer the telephone number to the Sometown exchange to indicate the destination of the call and this can be done by sending pairs of tones down the trunk (similar to the tones generated by a modern push-button telephone).

During the 1960s, the use of signaling tones gave rise to a rather strange phenomenon known as "phone phreaking".[11] Phone phreaking seems to have developed independently in a number of different places, but one of the early pioneers was a blind American schoolboy named Joe Engressia. At the age of eight, it is said that Engressia liked to dial non-working telephone numbers just to listen to the recording that said "This number is not in service". One day, he started whistling a tune while listening to a recorded announcement and was surprised to find that the line suddenly went dead. When he repeated the experiment, the result was the same. Full of curiosity, he rang the local phone company and asked them why this had happened. An engineer explained that the phenomenon was known as "talk-off" and that it occurred when a 2,600 Hz tone occurred by chance during the course of a telephone conversation.

Joe subsequently discovered that he could use this knowledge to obtain free telephone calls. In 1969, while a student at the University of South Florida, he was disciplined by the college after being caught whistling up free long-distance calls for fellow students. His exploits eventually led to him being arrested and prosecuted for "malicious mischief" after he featured prominently in a famous 1971 article on phone phreaking in *Esquire* magazine.

Joe Engressia had perfect pitch, which meant that he could whistle an accurate 2,600 Hz tone whenever he wanted. However, for the benefit of lesser mortals without perfect pitch (or those who could not whistle), electronic "blue box" tone generators soon started to appear. These boxes were held next to the telephone mouthpiece and enabled phone phreaks to generate the full range of

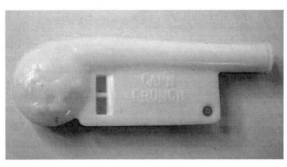

*Cap'n Crunch* whistle.

signaling tones that they needed. It was even discovered that the plastic whistle included as a free gift in packets of *Cap'n Crunch* breakfast cereal would produce the required 2,600 Hz tone.

Phone phreaks developed a number of techniques for setting up free telephone calls and they loved to demonstrate their expertise by setting up calls with long and devious routes. For example, a call might be established between two adjacent telephones with a route that threaded its way around the world, thereby enabling the speaker to listen to the sound of their own voice. Telephone conference calls were organised—free, of course—during which phreaks would exchange information about the inner workings of the telephone system and brag about their latest exploits.

A common phreaking technique that was used on North American networks worked as follows:

1. A toll-free (800) number belonging to a company in another city would be dialed in the normal way. Since this was a long-distance call, the local exchange would route the call to an available long-distance trunk.
2. Before the call was answered, the phone phreak would send a 2,600 Hz tone down the line using a blue box. This tone would be detected by the exchange at the far end of the trunk and would be interpreted as a signal to clear down the call. The local exchange would not detect the tone, so would assume that the original call was still going through.
3. The phone phreak would then switch off the 2,600 Hz tone. This would be interpreted by the remote exchange as the start of a new call.
4. The phone phreak could now use his blue box to issue dialing instructions to the remote exchange. The local exchange still believed that the call was going to the original toll-free number, so the subscriber would not be charged for the call.

Since this technique was illegal, calls of this type were often made from public phone boxes, but the fear of prosecution seems to have done little to deter dedicated phone phreaks. It was only with the introduction of digital telephone exchanges in the 1980s that phone phreaking finally came to an end. Digital exchanges are rather like computers and they communicate with each other in much the same way as computers communicate over the Internet. As digital data networks replaced analog tones for signaling between telephone exchanges, the opportunities for using blue boxes disappeared.

$$\star \quad \star \quad \star$$

At the start of this chapter, we asked how telephones could be connected together to form a network. We are now in a position to answer this question. Let's redraw Figure 11 for the telephone instead of the telegraph. Row A in Figure 17 illustrates a local call established between two subscribers connected to the same local exchange. Row B illustrates a call between two subscribers connected to different local exchanges, but where the two local exchanges share the same

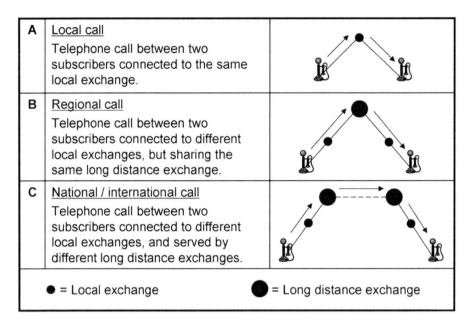

| A | Local call |
|---|---|
|   | Telephone call between two subscribers connected to the same local exchange. |
| B | Regional call |
|   | Telephone call between two subscribers connected to different local exchanges, but sharing the same long distance exchange. |
| C | National / international call |
|   | Telephone call between two subscribers connected to different local exchanges, and served by different long distance exchanges. |

● = Local exchange         ⬤ = Long distance exchange

**Figure 17.**   Telephone call routes between exchanges.

long-distance exchange. This is typically the case for regional calls. Row C illustrates a call between two subscribers connected to different local exchanges, where the two local exchanges are not connected to the same long-distance exchange.[12] This is typically the case for national and international calls.

This use of local and long-distance exchanges developed during the early years of the telephone network. Although the technology used in modern networks is completely different from the manual switchboards used in the nineteenth century, the basic architecture of a modern telephone network is remarkably similar. It is illustrated in Figure 18.

The access network provides a direct link from each subscriber to their local exchange. Traditional access networks consist mainly of long lengths of copper wire, while recently built access networks tend to include more exotic technologies such as radio and optical fiber.

The regional network provides the linkages between local and long-distance exchanges. Each local exchange is typically linked to at least two long-distance exchanges so that the failure of a single link cannot leave the local exchange isolated from the network.

The core network provides the high-capacity trunks that link the long-distance exchanges together. For a national telephone network, the number of long-distance exchanges is likely to be far smaller than the number of local exchanges, so it is often possible to provide a direct link from every long-distance exchange to every other long-distance exchange. This approach has the advantage that long-distance calls do not need to be switched at more than

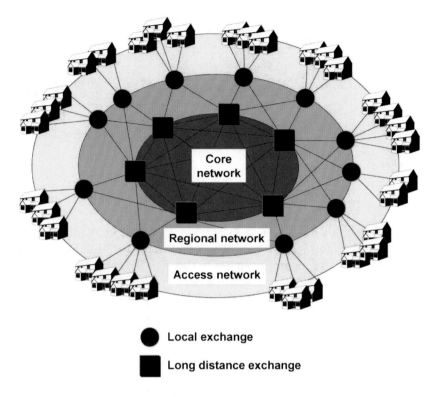

**Figure 18.** Network architecture.

two long-distance exchanges (unless the call terminates on a network belonging to a different operator).

\*　\*　\*

Although the telephone pioneers were able to learn many useful lessons from the development of the telegraph, their task was far from straightforward. In some cases, telegraph technologies could not be adapted to meet the needs of the emerging telephone network because of one fundamental difference: the telegraph was essentially a digital device, while the telephone was entirely analog.

To illustrate this point, let's consider the problem of signal attenuation on long-distance lines. During the early years of the telephone network, phone calls of more than a few dozen miles were not possible because no suitable form of amplification was available. The telegraph, on the other hand, could operate over links that stretched around the world. This was because, as we saw in Chapter 1, telegraph lines used relays to regenerate the signal at regular intervals.

Figure 19 illustrates how signal regeneration works. The left-hand side of the figure shows a telegraph signal representing the letter N (- ·). By the time this

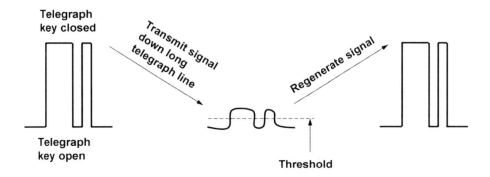

Figure 19.  Signal regeneration.

signal has traveled down a long length of wire, it is a distorted and considerably weaker version of its former self. However, if a suitable threshold is chosen, it is easy to detect when the incoming signal is above or below the threshold and a new and more powerful signal can be generated. This ability to clean up distorted signals is one of the key advantages of digital technologies.

Regeneration works because a Morse telegraph signal can take only two possible values (key open/key closed), while an analog waveform can take any value within a given range. In theory, a weak analog signal could be boosted by passing it through an amplifier, but the required technology did not exist in the nineteenth century. Furthermore, if the analog waveform has also been distorted by its passage down the telephone line, then it is normally impossible to recreate the original signal.

By about 1890, telephone engineers had managed to extend the range of telephone conversations to a few hundred miles. In 1906, Lee de Forest invented the Triode,[13] thereby making it possible to amplify signals electronically. The Bell system bought the rights to De Forest's patents in 1913, and by 1915 a transcontinental telephone service had been established across the United States. By the mid 1920s, long-distance lines had been extended to every part of the United States.

However, the telephone was still unable to challenge the dominance of the telegraph on really long-distance routes such as London to New York.[14] A trans-Atlantic telephony service using radio was launched in 1927, but it was not until the invention of submarine amplifiers[15] that trans-Atlantic telephony over cable finally became possible in 1956. A problem that had been solved on the telegraph network by the middle of the nineteenth century was not solved on the telephone network until the middle of the twentieth century!

As telephone usage started to grow in the late nineteenth century, the need arose to reduce the cost of installing long-distance telephone lines. As we saw in Chapter 2, attempts to carry two telegraph circuits on a single telegraph wire had started as early as 1853 and Emile Baudot had developed a sextuplex telegraph based upon a rotating "distributor" in 1874. This was a mechanical implementation of a technique that engineers would now call Time Division Multiplexing (or TDM). At about the same time, Alexander Graham Bell was also working on the problems of multiplexing telegraph circuits, but his approach was entirely different. His "harmonic telegraph" used a different frequency to carry each telegraph channel—a technique that would now be called Frequency Division Multiplexing (or FDM).

Time division multiplexers divide the capacity of a high-speed trunk circuit into time slices. If eight channels are sharing the trunk, then each channel is given access to the capacity of the trunk during every eighth time slice. So long as the trunk is transmitting information more than eight times faster than each channel, all of the channels will be able to operate satisfactorily. Frequency division multiplexers, on the other hand, divide the capacity of a high-speed trunk into frequency bands. Each telephone signal is modulated on to a different carrier frequency in much the same way as radio stations broadcast on separate frequencies. If eight channels are sharing the trunk and the band of frequencies carried by the trunk is more than eight times larger than the frequency band occupied by each channel, then all of the channels will be able to operate simultaneously over the same trunk.

As a general rule, TDM techniques are more suited to multiplexing digital traffic, while FDM techniques are more suited to analog traffic. FDM was the natural choice for the analog telephone network, but FDM could not be implemented using nineteenth-century electro-mechanical technology. As a result, multiplexing was not introduced on the telephone network until the 1930s—some 60 years after it had first appeared on the telegraph network. More recently, the introduction of digital technologies into the telephone network has enabled the use of TDM—a technique that appeared in Baudot's telegraph system of 1875.

Another problem faced by the telephone pioneers was that of electrical noise. Noise is the term used by engineers for any unwanted electrical signal, so the interference seen on a television screen when a vacuum cleaner is being used nearby would be described as noise. Once an analog signal has been contaminated by electrical noise, it is usually impossible to recover the original signal. For digital signals, on the other hand, it is normally possible to eliminate the effects of noise by periodically regenerating the signal in the manner described in Figure 19. For this reason, noise was never a significant problem on the telegraph network, but it proved to be a major challenge on the early telephone network.

Noise is not always a bad thing.[16] The low-level hiss that is heard during gaps in a telephone conversation would be regarded by most people as reassuring and phone companies receive complaints if they reduce it too much. However, early telephone users were subjected to something far more intrusive, as graphically described by Herbert N. Casson in 1910:

> "Noises! Such a jangle of meaningless noises had never been heard by human ears. There were spluttering and bubbling, jerking and

rasping, whistling and screaming. There were the rustling of leaves, the croaking of frogs, the hissing of steam, and the flapping of birds' wings. There were clicks from telegraph wires, scraps of talk from other telephones, and curious little squeals that were unlike any known sound. The lines running east and west were noisier than the lines running north and south. The night was noisier than the day, and at the ghostly hour of midnight, for what strange reason no one knows, the babel was at its height."[17]

After considerable experimentation, it was found that the noise was caused by the use of an earth return. As we saw in Chapter 2, earth returns were widely used on the telegraph network because they halved the amount of wire required to build a circuit. However, telephone engineers were eventually forced to the conclusion that the only way to build a telephone network free from excessive noise was to eliminate the earth return and to carry both halves of each circuit on wire.[18] This considerably increased the cabling costs and there was worse to come—when amplification was finally introduced on long-distance lines, the amplifier could only operate in one direction, so a separate pair of wires was required for each direction of transmission. Beyond that, additional wires were sometimes required to carry signaling. While a telegraph link could be implemented using a single piece of wire, a long-distance telephone line required at least four.

The long list of problems faced on analog telephone networks suggests that introducing digital technologies into telephone networks might bring considerable benefits. As we shall discover in the next chapter, it was not until the second half of the twentieth century that it became technically and commercially possible to do this. However, by the beginning of the twenty-first century, digital technologies developed for the Internet would be driving some of the most rapid evolution that the telephone network had ever seen.

# 5    Going digital

At the end of the last chapter, we discussed how the digital nature of the telegraph network enabled it to avoid some of the problems encountered on the analog telephone network. In this chapter, we will explore how and why the analog telephone network finally converted to digital.

Speech is an analog waveform. As we saw in Chapter 3, the various sounds that make up human speech are carried through the air by pressure fluctuations. When these pressure fluctuations are detected by a telephone microphone, they are converted into a fluctuating electrical signal—or a "voice-shaped current", as Alexander Graham Bell called it. If we are to carry this electrical signal over a digital network, then we need to convert it into a stream of digits that can be used at the receiver to recreate the electrical signal.

So, why would we want to go to the trouble and expense of converting an analog signal to digital and then back to analog again? Some of the principal reasons are listed below.

- Once signals as diverse as voice, video and data have all been converted to streams of bits, they no longer need to be carried on separate networks. This reduces networking costs and creates opportunities for new multimedia services.
- Digital compression techniques can be used to squeeze more information into the same amount of bandwidth, thereby further reducing costs.
- Digital signals can emerge sparkly and new from a transmission system that introduces significant degradations (such as attenuation and noise).[1]
- In spite of the quality of digital transmission systems, occasional errors can occur. However, by adding some additional bits to a digital signal, it is possible for the receiver to check for errors and to automatically request re-transmission if an error is detected. In some cases, it is even possible to correct errors without the need for re-transmission. Clearly, error detection and correction are essential for applications such as financial transactions, in which a single corrupted bit could prove to be extremely expensive.
- Powerful digital encryption techniques can be used to provide security. This allows sensitive voice conversations to take place over insecure radio links.[2]
- Digital technologies have been driven forward during the last few decades by advances in the computer industry and by the availability of highly integrated "silicon chips". Telephone networks have been able to exploit these advances to improve reliability and reduce costs.
- Cheap digital technology enables software-based intelligence to be

A. Wheen, *From Dot-Dash to Dot.com: How Modern Telecommunications Evolved from the Telegraph to the Internet*, Springer Praxis Books, DOI 10.1007/978-1-4419-6760-2_6,
© Springer Science+Business Media, LLC 2011

distributed to remote corners of the network. This can make it simpler and cheaper to identify and rectify faults when they occur.

● As network software becomes increasingly sophisticated, it can provide a wider range of information and services.

This is by no means a comprehensive list. Clearly, going digital is a *very* good idea!

⋆   ⋆   ⋆

So, how would we set about converting an analog waveform to digital? This is a subject that can become frighteningly mathematical when it crops up in engineering degree courses, but we are going to restrict ourselves to the basic principles. Please don't be frightened!

Let's start with a simple analog waveform, such as the one shown in Figure 20.

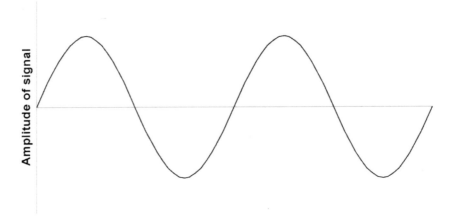

**Figure 20.**   Simple analog waveform.

This is a graph of an audio tone. If the tone is being produced by a loudspeaker or a telephone earpiece, then the vertical axis would represent pressure fluctuations in the air. However, let's assume instead that the tone is being carried electrically on a wire and that the vertical axis represents voltage fluctuations. The horizontal axis represents time, so the graph illustrates the shape of the waveform over a short period of time.

Now, let's assume that we measure the voltage of this waveform at equally spaced time intervals. These measurements are indicated by the dots in Figure 21. Each measurement produces a number that represents the voltage of the waveform at that particular moment in time, so a sequence of measurements produces a sequence of numbers. These numbers can be thought of as a digital representation of the waveform. If we were to transmit these numbers to someone at a remote location, then they could convert each number back to its corresponding voltage and regenerate an analog waveform similar to the one

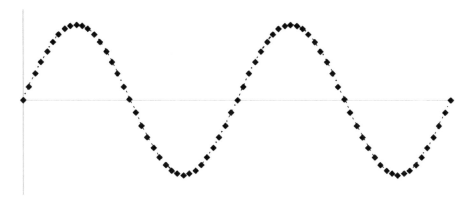

**Figure 21.** Voltage measurements.

that we had to begin with. It might not be absolutely identical, of course, but the diagram suggests that an accurate reconstruction of the waveform should be possible.

Now, let's suppose that we halve the number of samples that we collect per second, as illustrated in Figure 22.

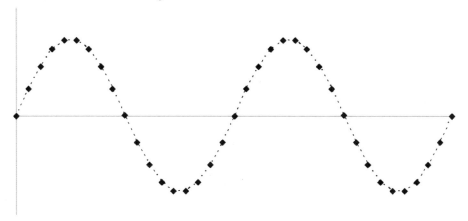

**Figure 22.** Voltage measurements with wider spacing.

The time between each sample is now twice as long, but the waveform is still being sampled sufficiently frequently to enable us to recreate it pretty accurately from the sequence of numbers. Furthermore, because we have halved the rate at which we are taking samples, we have also halved the bit rate required to transmit the digital signal. This is a good thing, because it means that the signal can be carried more cheaply across a network.

If halving the sample rate reduces the cost of transmitting the signal, then why not halve it again? Not surprisingly, engineers have investigated how far this process of sample rate reduction can be taken before the original waveform can no longer be satisfactorily reproduced at the receiver.

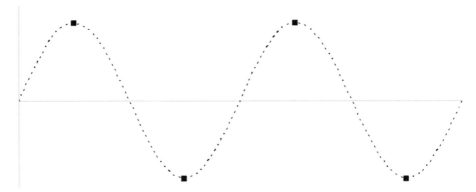

**Figure 23.** Waveform sampled at twice its highest frequency.

The answer, it turns out, is that the waveform must be sampled more than twice as fast as its highest frequency. In other words, if we are sampling a 2 KHz tone, then we would need to sample it more than 4,000 times per second to enable the receiver to reproduce it accurately. This limit is referred to as the "Nyquist Criterion".

Figure 23 illustrates the situation in which a waveform is sampled at exactly twice its highest frequency. However, if the samples in the diagram were to be delayed by a quarter of a cycle (i.e. moved slightly to the right), then they would all occur at moments when the waveform was crossing the horizontal axis of the graph and all the sample values would be zero. In this situation, the receiver would have no information about the amplitude of the waveform. To avoid this problem, the Nyquist Criterion states that the sample rate must be *more than* twice the highest frequency contained in the waveform.

Figure 23 also raises another issue. Given the same set of samples, there are many other waveforms that would fit equally well, such as the one illustrated in Figure 24. How can the receiver be certain what the original waveform really looked like?

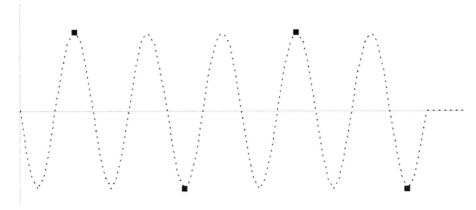

**Figure 24.** Fitting an alternative waveform to the same samples.

The answer lies in the fact that the waveform in Figure 24 is three times the frequency of the waveform in Figure 23, so it needs to be sampled three times more frequently if it is to be reproduced accurately. We can imagine many other waveforms that would fit the sample points shown above, but they would all contain frequencies that are too high to meet the Nyquist Criterion. There is, in fact, only one waveform that fits the sample points AND meets the Nyquist Criterion.

This phenomenon of multiple waveforms fitting the same set of sample points is known as "Aliasing"[3] and it occurs whenever the sample rate is too low. A classic illustration of this effect can be seen in western movies when a carriage rolls into town and draws to a halt. As it slows down, there is a period of time when the spokes of the wheels appear to rotate backwards. This occurs because a film is not really a continuously moving picture—it is actually a sequence of photographs ("frames") that are displayed in such quick succession that the viewer's brain is fooled into perceiving continuous movement. The spokes on a carriage can move quite rapidly, while a movie camera typically records only about 25 frames per second. This sample rate is too slow to meet the Nyquist Criterion and an alias is created in which the carriage wheels appear to rotate backwards.

A simple way of preventing aliasing problems on digital networks is to filter the analog waveform before it is sampled to remove any frequencies that are too high to meet the Nyquist Criterion. In a typical telephone network, each channel is filtered to remove speech components above about 3.4 KHz and the resulting waveform is then sampled at 8 KHz (i.e. 8,000 times per second).[4] The channel is also filtered to remove frequency components below about 300 Hz to eliminate mains hum and other undesirable signals that might leak into the telephone wiring. Experiments carried out many years ago established that this 300 Hz to 3.4 KHz frequency range is sufficient to carry "telephone quality" speech.

As mentioned previously, reducing the bit rate of a digital circuit is a good thing—at least from the point of view of the telephone company—because it reduces the cost of carrying a telephone conversation across a network. One way of reducing the bit rate of a digital signal is to reduce the number of samples that we take per second from the analog waveform; however, as we have just seen, there are limits to how far we can take this process without encountering aliasing problems. Another way to reduce the bit rate is to reduce the number of bits used to represent each sample and this brings us to another potentially scary subject: quantization. As before, we will restrict our discussion to the basic principles. Please be brave!

Let's assume that our analog waveform can take any value in the range 0–10 V. If we can measure the voltage of each analog sample to an accuracy of 1 V, then the digital representation of each sample is restricted to one of 10 possible values and can be represented unambiguously by a single decimal digit (Table 3).

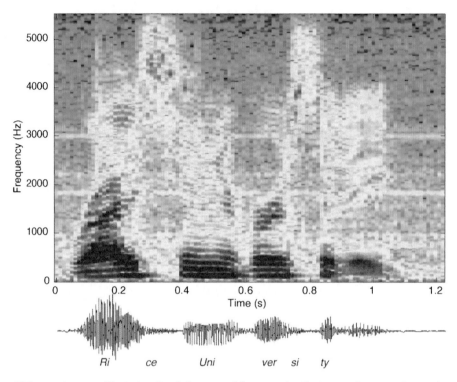

This spectrogram illustrates the full range of frequencies that occur in normal speech. The lower part of the spectrogram shows the speech waveform generated by someone saying "Rice University". The upper part shows the amount of energy in the speech at different frequencies (red indicates a high level of energy, while blue indicates a low level of energy). As can be seen, speech contains frequency components above 3.4 KHz, but these components occur mainly in unvoiced sounds that do not involve the vocal chords (e.g. the "ce" at the end of "Rice", and the "si" in "University"). This explains why it can be extremely difficult to differentiate between sounds such as "f" and "s" on the telephone—most of the energy in these sounds is above 3.4 KHz. For this reason, the Skype Internet telephony service often carries an extended range of frequencies by sampling at 16 KHz rather than 8 KHz. In digital systems in which very high quality is required, an even wider range of frequencies needs to be encoded; young people can hear up to about 20 KHz, so audio CDs use a sampling rate of 44.1 KHz.

**Table 3.**   Representing sample measurement with a single digit.

| Voltage (V) | 0–1 | 1–2 | 2–3 | 3–4 | 4–5 | 5–6 | 6–7 | 7–8 | 8–9 | 9–10 |
|---|---|---|---|---|---|---|---|---|---|---|
| Digit | 0 | 1 | 2 | 3 | 4 | 5 | 6 | 7 | 8 | 9 |

However, since we can only measure to the nearest volt, a sample of 3.24 V would be represented by the same decimal digit as a sample of 3.86 V. This process of forcing a measurement to adopt one of a restricted number of possible values is known as quantization.

If an accuracy of 1 V is not sufficient, then we could use a device that measures each sample to an accuracy of 0.1 V. Each measurement would now have 100 possible values, so we would need two decimal digits to represent the value of each sample (Table 4).

**Table 4.** Representing sample measurement with two digits.

| Voltage (V) | 0–0.1 | 0.1–0.2 | 0.2–0.3 | 0.3–0.4 | . . . | 9.8–9.9 | 9.9–10 |
|---|---|---|---|---|---|---|---|
| Digits | 00 | 01 | 02 | 03 | . . . | 98 | 99 |

We can (at least in theory) follow this line of reasoning to any level of accuracy that we desire. As we improve the accuracy of each measurement, we improve the accuracy of the analog waveform that can be recreated at the receiver. However, increasing the accuracy of each measurement also increases the amount of digital information that needs to be transmitted, and this has implications for the cost of operating the network. A point is eventually reached at which the voltage measurements are sufficiently accurate to produce acceptable speech and the benefits of additional accuracy are not justified by the additional costs.

Up to this point, we have talked about the digital signal in terms of decimal numbers. In practice, modern telephone networks use a binary representation of the analog signal rather than a decimal one. Binary digits (usually referred to as "bits") are much more convenient than decimal digits to handle electronically because they can have only two possible values: 0 and 1. For this reason, the internal arithmetic operations that go on inside your PC also use binary rather than decimal arithmetic.

Let's summarize what we have learned so far. The process of converting from analog to digital[5] consists of three principal stages, as illustrated in Figure 25.

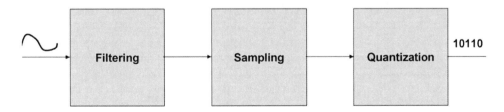

**Figure 25.** Analog-to-digital conversion process.

*Filtering* sets an upper limit on the frequencies that can be contained in the analog waveform. This is necessary to meet the Nyquist Criterion and thereby prevent aliasing. In the telephone network, filtering is also used to eliminate some undesirable low-frequency components from the speech signal.

*Sampling* captures the value of the filtered analog waveform at regular intervals. In the case of the telephone network, the speech waveform is sampled 8,000 times per second. The analog waveform between these sample points is discarded.

*Quantization* is the process of measuring the value of each sample and converting it to the nearest digital equivalent. A quantizer that produces an output containing a large number of decimal places will provide an extremely accurate digital representation of the analog signal. However, a high bit rate will be required to transmit the digital information across the telephone network and this has cost implications.

Having considered how an analog speech signal can be converted to digital, we now need to consider how the digital signal can be converted back to analog. This process is illustrated in Figure 26.

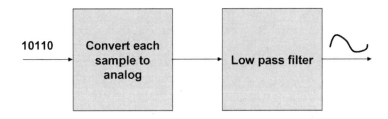

**Figure 26.** Digital-to-analog conversion process.

Let's assume that the incoming digital signal contains 8,000 sample values per second. Every 1/8,000 of a second (i.e. every 125 microseconds), a new sample value is applied to the input of a digital-to-analog converter, which generates an analog output voltage corresponding to the digital input value. This produces a stepped waveform, as illustrated in Figure 27.

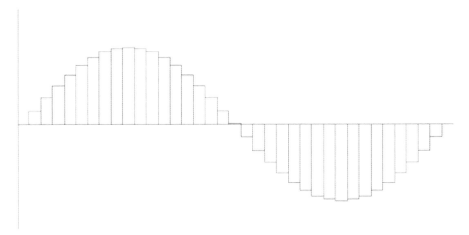

**Figure 27.** Output from digital-to-analog conversion.

This stepped waveform is obviously not identical to the smooth waveform that we had at the transmitter, but we can make it smoother by passing it through a low pass filter. Sadly, it still won't be identical to the original transmitted waveform because information was lost during the quantization process. The differences between the transmitted and received waveforms can be regarded as low-level noise—or "quantization noise"—that has been super-imposed on the original signal. Quantization noise is the price that we pay for restricting the number of output levels available to the quantizer at the transmitter.

People who are unfamiliar with digital technology are often bothered by the concept of quantization noise. This, they argue, is an entirely avoidable source of noise that does not occur in analog systems. Whilst this is undeniably true, it is also missing the point. In digital systems, quantization noise is effectively the *only* source of noise that we need to worry about and we can make it as small as we like by increasing the accuracy of the quantizer. As we saw in the previous chapter, we are effectively immune from other sources of noise once we have entered the digital domain. In contrast to this, any noise that manages to contaminate an analog signal is difficult or impossible to remove. Quantization noise is a very small price to pay for the benefits that digital technology brings.

* * *

One of the key moments in the development of digital telephony occurred in 1928 when Harry Nyquist published a paper called—rather unpromisingly— "Certain Topics in Telegraph Transmission Theory". This introduced the fundamentally important concept that has since become known as the "Nyquist Criterion" and was introduced earlier in this chapter. Harry Nyquist was born in Nilsby, Sweden, in 1889. He emigrated to the United States in 1907 at the suggestion of a teacher who had been impressed by his intellectual abilities. After degrees at the University of North Dakota and a PhD in Physics at Yale University, Nyquist joined the American Telephone and Telegraph (AT&T) company in 1917. His department was merged into the Bell Telephone Laboratories in 1934 and he remained there until his retirement in 1954.

One of Nyquist's first assignments at AT&T concerned the development of a facsimile machine for use over telephone lines. He subsequently extended the work of J. B. Johnson and arrived at a mathematical explanation for thermal noise (now known as Johnson–Nyquist noise). He also did important work on the stability of negative feedback amplifiers, resulting in the Nyquist Stability Theorem. During his 37 years with AT&T/Bell, Nyquist was responsible for a total of 138 US patents, but it was probably his work on sampling that has had the greatest impact on the modern world.

We have seen that there are three key stages when converting an analog signal to digital: filtering, sampling and quantization. Nyquist's 1928 paper addressed the issues of filtering and sampling, but it did not address quantization. It was not until 1937 that all three operations were finally brought together by Alec

Reeves in a digital coding scheme known as Pulse Code Modulation (PCM). Alec Reeves was born in March 1902 in Redhill, Surrey. After studying engineering at Imperial College, London, he joined International Western Electric in 1923. Two years later, the company was taken over by ITT and Reeves went to work at ITT's laboratory in Paris. It was while he was in Paris that he filed his PCM patent.

The technical principles of PCM were actually described at the start of this chapter, although we refrained from using such a scary name at that stage. After an analog waveform has been filtered and sampled, each sample is then quantized so that it can be represented by a binary number. Since the speech waveform is transmitted down the line as a series of ones and zeros, it can (at least in principle) be regenerated even if it has been severely attenuated or contaminated with noise. This can be done using essentially the same techniques that were used to regenerate telegraph signals.

Modern PCM systems transmit speech at 64 Kbit/second. Reeves recognized that the bit rate required to transmit a group of PCM channels would be a significant challenge for the long-distance transmission technology available at the time, but he assumed (correctly) that it would eventually become economically viable. He built the world's first PCM multiplexer in 1937 to demonstrate the principles involved, but its valve-based technology was expensive and unreliable. It was not until the development of solid-state semiconductor devices in the 1960s that digital multiplexers finally started to appear on telephone trunk routes. Since then, the practical applications of PCM have stretched far beyond the bounds of telephone networks. Although they may not know it, users of digital devices such as radios, televisions, mobile phones, CDs and DVDs owe Alec Reeves a debt of gratitude.[6]

Reeves left France in some haste when the Germans invaded in 1940. Back in Britain, he joined Scientific Intelligence and was soon working on the problems of night navigation for bombing aircraft. He proposed a system in which a pilot flew around an arc centered on one base station and then dropped his bombs when he crossed an arc transmitted from a second base station. The distance of the aircraft from each base station was determined by transmitting a radio pulse from the base station and then measuring the time taken for it to be returned by the aircraft. The system provided the pilot with an audible warning if he was deviating from the correct path and the sound that this made gave the system its name: Oboe.

The accuracy of Oboe was remarkable and was not matched until the development of satellite navigation and laser guidance systems many years later. It was used by the RAF for over 9,000 raids and it improved the effectiveness of allied bombing to such an extent that it had a very real impact on the course of the Second World War. Later in the war, Oboe was also used for dropping food canisters to Dutch resistance fighters living under German occupation.

After the war, Reeves worked at STL in Harlow (part of ITT) on ways to improve the capacity and reliability of communications systems. In the late 1960s, he provided the inspiration for Charles Kao and George Hockham in their groundbreaking research on fiber optics. During his life, he was awarded over 100 patents.

As with many scientists and inventors of his stature, Reeves had an unconventional side. He had a keen interest in spiritualism and the paranormal and conducted complex experiments to investigate communication with the dead. He believed that he was in regular communication with the nineteenth-century scientist Michael Faraday and with an American Indian called Red Cloud. Alec Reeves died in 1971.

Following Reeves's 1937 PCM patent, the next major milestone in the development of digital communication was the publication in 1948 of a paper entitled "A Mathematical Theory of Communications" by Claude E. Shannon. This paper established the science of Information Theory and created a terminology and framework that are still in use today. It is no exaggeration to state that modern digital technology is built on foundations that were established by Shannon.

Claude Shannon was born in 1916 in Petoskey, Michigan. Inspired by his hero, Thomas Edison—whom he later learned was a distant cousin—the young Shannon showed a talent for constructing models and gadgets. He worked for a time as a telegram messenger for Western Union and built a private telegraph system to a friend's house half a mile away. He graduated from the University of Michigan in 1936 with twin BSc degrees in Electrical Engineering and Mathematics and then moved to the Massachusetts Institute of Technology (MIT) to study for a Master's degree. Shannon completed his PhD thesis on population genetics at MIT in 1940. In 1941, he joined Bell Laboratories and spent the rest of the Second World War working on fire control systems for anti-aircraft guns.

Shannon was fascinated by the form of algebra that had been developed by a nineteenth-century English philosopher called George Boole. He recognized that the two states of a Boolean variable (0, 1) could be mapped directly to the two states of a telephone relay (open, closed). Shannon's MSc thesis, "A Symbolic Analysis of Relay and Switching Circuits", showed how Boolean logic and binary arithmetic could be used to simplify the arrangement of relays used in telephone exchanges and how—conversely—arrangements of relays could be used to solve problems in Boolean algebra. This new approach turned out to be fundamentally important for the design of digital circuits and lies at the heart of modern digital computers.

In 1948, Shannon published his seminal work, "A Mathematical Theory of Communications", in the *Bell System Technical Journal*. This paper introduced the concept that all forms of information (voice, video, text, etc.) can be represented mathematically by a sequence of ones and zeros. He showed that these ones and zeros can be treated separately from the communication channel that carries them, thereby dismissing the idea that telegrams and telephone calls have to be carried on separate networks. This paper introduced the word "bit" for the first time and demonstrated how the maximum capacity of a communication channel can be measured in terms of bits per second. It was Shannon who established the bit as the fundamental element of digital communications.

Shannon's paper was heavily influenced by the work of Harry Nyquist, who was a colleague of his at Bell Laboratories. While Nyquist was the first person to provide a clear statement of the Nyquist Criterion, Shannon was the first to provide the mathematical justification.

In 1956, Shannon became a Visiting Professor of Communication Sciences and Mathematics at the Massachusetts Institute of Technology (MIT) and, 2 years later, he became Donner Professor of Science. He developed a keen interest in artificial intelligence and developed a mechanical mouse (complete with copper whiskers) that could find its way to the centre of a maze. He also carried out some early work on programming digital computers to play chess.

It is a measure of Shannon's intellectual stature that many of his most brilliant colleagues at MIT and Bell Laboratories considered him to be a genius. Comparisons were drawn with Einstein and although Shannon never achieved Einstein's "mad professor" appearance, his behavior was sometimes delightfully erratic. He would often be seen riding a unicycle down the corridors at MIT and would sometimes enhance the effect by juggling at the same time. He later invented a two-seater unicycle (although it appears that nobody was anxious to share it with him) and a unicycle with an off-center hub that bobbed up and down. Other inventions attributed to him include a motorized pogo stick, a flame-throwing trumpet and a rocket-powered frisbee.

Sadly, the man who did so much to unleash the marvels of the digital revolution became increasingly unaware of its consequences as his mind was ravaged by Alzheimer's disease. He spent his last few years in a Massachusetts nursing home and died in 2001.

*  *  *

Most telephone networks built since the early 1990s have been entirely digital from day one, but the process of converting long-established analog networks to digital is by no means straightforward. To begin with, the level of investment in a large analog network is so great that it is not financially viable to replace the network over a short period of time. Furthermore, the need to maintain services to customers means that it is completely impossible to shut down an existing network and start again. For these reasons, the process of telephone network digitization has had to proceed rather slowly and the world's networks still contain plenty of analog equipment.

Telephone networks are made up of many component parts (local exchanges, tandem exchanges, trunk circuits, subscriber lines, etc.) and it is possible to convert one part to digital without affecting the rest of the network. As an example, an analog telephone exchange can be replaced by a digital exchange with analog interfaces. When a trunk circuit is subsequently upgraded to digital, the corresponding analog interface on the telephone exchange can be replaced with a digital interface. This step-by-step approach to digitization means that digital "islands" will start to develop within an analog network. Telephone calls can pass through a number of these digital islands as they make their way across the network, with each one introducing an additional conversion from analog to digital and back again. Too many of these conversions can degrade the quality of the telephone call, so network planners need to be careful to ensure that this does not happen. As the digital islands start to merge to form larger digital areas within the network, the number of conversions between analog and digital starts to reduce. When this process of digitization has been taken to its logical conclusion, speech is converted to digital within the customer's telephone handset and remains digital until it arrives at the receiving handset. This level of digitization has already been achieved on GSM mobile networks. However, even in developed countries, many non-mobile telephone calls are still carried in analog form between the customer's premises and the local exchange.

One way of transmitting digital traffic over analog parts of the network is to use a device called a voiceband modem (short for MOdulator/DEModulator), as illustrated in Figure 28.

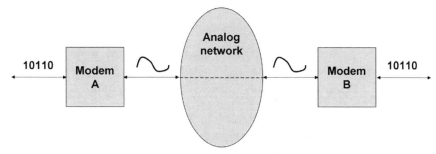

**Figure 28.** Using modems to transmit data over an analog network.

A digital signal traveling from left to right arrives at Modem A and is converted into an analog signal that fits within the 300 Hz to 3.4 KHz frequency range that the telephone network can carry. When this analog signal arrives at Modem B, it is converted back to digital. Since modems (and telephone lines) are bi-directional, data can also be sent in the opposite direction at the same time.

Now, let's consider how the situation would change if the analog network has been partially converted to digital. The analog network would now contain islands of digital equipment and the modem signal might have to pass through a number of these islands as it crosses the network. This is illustrated in Figure 29.

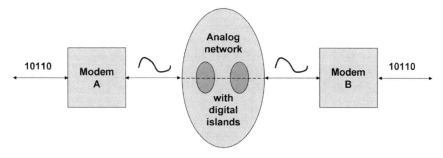

**Figure 29.** Analog network with digital islands.

The signal probably started life in a digital computer and was then converted to an analog data signal by a modem before being sampled and digitized by a digital voice channel. This will normally work satisfactorily, although it is a highly inefficient way of transmitting information.

The first modems started to appear in the early 1960s and operated at the dizzying speed of 300 bit/second. The speed was soon increased to 1.2 Kbit/second and it had reached 9.6 Kbit/second by 1984. Ten years later, it had reached 28.8 Kbit/second. The current limit is 56 Kbit/second, but this is still hundreds of times slower than a typical office data network and illustrates one of the fundamental problems with modems—they are far too slow for most modern applications.

A second problem with modems arises from a key difference between voice and data: voice can be carried satisfactorily using a steady bit rate (such as 64 Kbit/second), while data tends to be much more "bursty". As an example, if you are surfing the web and you type in the address of a website, many kilobytes of data will have to be transferred to your PC within a few seconds, but it may then be another 10 minutes before you request another web page. The fixed bit rate provided by a modem link is far too slow when data is actually required, but the bits are still flowing—and thereby incurring telephone charges—when data is *not* required.

Although modems have long since been replaced by high-speed digital circuits and data networks in most business applications, they still exist inside devices such as fax machines, credit card payment machines and satellite TV set-top boxes. Modems also found a new lease of life providing dial-up access to the Internet, although the growth of broadband is now displacing them from this application. In spite of this, it would be unwise to predict that the modem will completely disappear in the near future. To paraphrase Mark Twain, reports of the death of the modem have often been greatly exaggerated.

Many years ago, the telephone network replaced the telegraph network. Although people continued to send telegrams,[7] those telegrams were transmitted over the telephone network and the costs associated with building and operating two completely separate networks were eliminated.[8] Today, the Internet and the telephone network are operating in parallel, but convergence is starting to occur in the form of "next-generation networks". The dramatic growth in data communication means that voice is becoming an increasingly insignificant proportion of total network traffic. As a result, new networks are adopting designs that are much closer to the Internet than to traditional telephone networks. These radical developments will be discussed in more detail in Chapter 13.

# 6

# A bit of wet string

We learned in Chapter 4 how regional and core networks enable long-distance telephone calls to be set up between different parts of the country. Transmission systems provide the trunks that carry these calls from one telephone exchange to the next. In popular mythology, this part of the network is often characterized as being little more than a bit of wet string, but it is actually an area in which some very sophisticated technology can be found. This chapter considers how these links have evolved from wires carrying a single telephone call into fiber optic cables carrying voice, data and video traffic in mind-boggling quantities.

Before we start, it is useful to be clear about what we are looking for. When discussing transmission systems, how can we decide that one is "better" than another? Fortunately, the answer to this question can be stated rather concisely: in simple terms, an ideal transmission system would have zero *attenuation*, zero *delay* and infinite *bandwidth*. And, for good measure, it would cost nothing to install or to operate! Clearly, this is an ideal that we can never achieve in practice, but it is important to understand our objective as we strive to get as close to it as we can. So let's start by clarifying what we mean by attenuation, delay and bandwidth.

The problem of signal *attenuation* is one that has troubled communication engineers since the earliest days of the telegraph and we first encountered it in Chapter 1. In simple terms, attenuation measures the proportion of the signal that is lost as it is transmitted from one place to another. As we saw in Chapter 4, it is important to regenerate a signal before attenuation becomes too severe.[1] Since regeneration costs money, attenuation is definitely something that we want to minimize.

The *delay* introduced by a transmission system represents the amount of time that elapses between a signal entering at one end of the system and emerging at the other. Transmission delay is caused partly by the equipment at each end of the link and partly by the repeaters at intermediate points in the link. Amazingly, it is also caused by the speed at which an electrical signal can propagate down a cable—or the speed at which light can propagate down an optical fiber. For a typical telephone call, these delays add up to about 1 millisecond (one-thousandth of a second) per 100 miles. This may not seem like much, but delays can easily reach the point on long-distance lines at which they start to cause noticeable echo on the line. This is particularly true for calls that are carried by communications satellites, where the delay is quite long enough to make telephone conversations difficult.[2]

*Bandwidth* measures the range of frequencies that the transmission system can

A. Wheen, *From Dot-Dash to Dot.com: How Modern Telecommunications Evolved from the Telegraph to the Internet*, Springer Praxis Books, DOI 10.1007/978-1-4419-6760-2_7,
© Springer Science+Business Media, LLC 2011

carry. To take a simple example, a channel that can transmit frequencies in the range 300 Hz to 3.4 KHz has a smaller bandwidth than a channel that can transmit frequencies between 300 Hz and 7 KHz. In the digital domain, a transmission system with infinite bandwidth would be able to transmit an infinite number of bits per second.

Now that we know what we are looking for, let's consider how we might provide a trunk connection between two telephone exchanges. There are a number of possible options:

- copper cable;
- radio;
- satellite;
- fiber optic cable.

For the telephone pioneers in the late 1870s, it is arguable that none of these options was available. Radio was unknown, although some amazing mathematical predictions by James Clark Maxwell had suggested that radio waves ought to exist and experiments by David Hughes[3] had indicated that communication without wires might be possible. Radio did not become a viable form of communication until the early twentieth century, and communication satellites did not emerge from the realms of science fiction until the 1960s. Fiber optics took even longer to appear, in spite of the fact that Alexander Graham Bell had developed a telephone that transmitted speech using a beam of light as early as 1880.

Surprising as it may seem, even copper cables were not available to the telephone pioneers. Early telephone wires were made of galvanized iron or steel in spite of the fact that these metals were poor conductors of electricity. The ideal telephone wire would have been made from either silver or copper, but silver was much too expensive and copper was too weak to support its own weight. However, in 1877, Thomas Doolittle developed a process to increase the tensile strength of copper wire by drawing it through a series of dies, thereby making it strong enough for use on telephone lines. The new wire was deployed in various parts of the United States to assess its performance in different climates and it proved to be a great success. Writing in 1910, Herbert Casson observed that:

> "... there has been little trouble with copper wire, except its price. It was four times as good as iron wire, and four times as expensive. Every mile of it, doubled, weighed two hundred pounds and cost thirty dollars. On the long lines, where it had to be as thick as a lead pencil,[4] the expense seemed to be ruinously great. When the first pair of wires was strung between New York and Chicago, for instance, it was found to weigh 870,000 pounds—a full load for a twenty-two-car freight train; and the cost of the bare metal was $130,000. So enormous has been the use of copper wire since then by the telephone companies, that fully one-fourth of all the capital invested in the telephone has gone to the owners of the copper mines."[5]

Although the problem of manufacturing copper telephone cables had been solved, a suitable insulating coating for the cable had still not been found. Cable insulation had been a major headache for the telegraph pioneers, as illustrated by Samuel Morse's difficulties on the Washington-to-Baltimore link (see Chapter 1). The problem had eventually been circumvented by suspending bare wires from glass insulators mounted on telegraph poles and this "open wire" technique had later been adopted for the early telephone network. However, water and dust particles soon reduced the effectiveness of the glass insulators by creating electrical leakage paths. Furthermore, overhead wires were vulnerable to lightening strikes and the need to maintain an adequate physical spacing between the wires created major problems in congested city centers. The solution to all of these problems would be to bury the cables underground, but damp soil conducts electricity rather too well to allow un-insulated wires to be buried.

This drove the continuing search for a suitable method of insulating cables and, in 1891, a dry core paper-insulated telephone cable was installed between Pipewellgate and Deckham in Gateshead-on-Tyne. This type of cable was soon widely used because the lead/tin alloy used for its outer sheath provided adequate water-proofing. After further developments, AT&T installed its first underground cable between Philadelphia and Washington, DC, in 1912. The discovery of polyethylene in 1933 finally provided a really effective form of cable insulation that could be used underground, thereby eliminating the need for open wire transmission systems.

*  *  *

To the layman, one piece of wire is much like another. Current will flow if you connect a battery at one end and a bulb at the other and it appears that the bulb illuminates at the instant at which the battery is connected. If, however, the wire is several miles long, we start to discover that things are actually much more complex than this. When an electrical signal propagates down an open wire transmission line, all sorts of horrible things start to happen.

Hans Christian Oersted showed in 1819 that a magnetic compass needle would move if placed near a wire carrying an electric current. This established for the first time that there is a link between electricity and magnetism. As a result of this effect, the electric current flowing in an open wire transmission line will generate a magnetic field surrounding the line. Since the current will be fluctuating in response to the voice signal that the line is carrying, corresponding fluctuations will also occur in the magnetic field. Faraday established in 1831 that a fluctuating magnetic field induces electric currents in nearby conductors, so currents will be induced in any metallic object in the vicinity of the telephone wire. As a result of these and other effects, transmission lines behave entirely differently from short pieces of wire and this places serious limitations on their ability to carry telephony traffic.

By the 1920s, the limited capacity of cables and open wire systems was becoming a major problem and the anticipated growth in telephone usage led

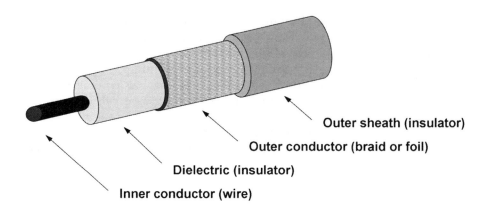

Outer sheath (insulator)

Outer conductor (braid or foil)

Dielectric (insulator)

Inner conductor (wire)

**Figure 30.** Coaxial cable.

engineers to search for a new form of cable that would be able to carry higher volumes of traffic on long-haul trunk circuits. In 1929, AT&T engineers Lloyd Espenschied and Herman Affel developed a "coaxial cable" consisting of two concentric cylinders of conducting material separated by an insulator (or "dielectric"). A cross-section through a coaxial cable is illustrated in Figure 30.

The outer conductor, which typically consists of wire braid or metal foil, is physically and electrically insulated from the outside world by a protective outer sheath. Inductive losses are significantly reduced because the electrical energy is confined within the dielectric. The outer conductor also acts as an electrical shield, thereby protecting the cable from external interference.

In November 1936, the first voice transmission was made over an experimental coaxial cable system installed between New York and Philadelphia. In the same year, the British Post Office installed a coaxial cable system that carried 40 telephone channels between London and Birmingham, with Frequency Division Multiplexing being used to enable multiple simultaneous calls on each cable. In the following year, a pair of submarine coaxial cables (a four-channel system and a 12-channel system) was laid between the United Kingdom and Holland. Long-distance coaxial cable systems were introduced in the United States in 1946. Early American systems could carry 600 telephone channels per cable pair and this number had increased to 13,200 channels per cable pair by the time that development of coaxial cable systems finally came to an end in the late 1970s.

The first trans-Atlantic telephone cable (TAT1) was based on coaxial cable and it could carry up to 36 simultaneous telephone calls. It may seem surprising that this cable did not appear until 1956—90 years after the first successful trans-Atlantic telegraph cable—but the problems of providing electronic signal amplification in the depths of the ocean were very real. However, people had been making telephone calls across the Atlantic for almost 30 years before the arrival of TAT1, using another important transmission technology: radio.

\*   \*   \*

The question of who actually invented radio communication continues to excite passions to this day. While Guglielmo Marconi is generally considered to be the father of radio, other claimants to this title include Nikola Tesla, Alexander Popov, Oliver Lodge, Reginald Fessenden, Heinrich Hertz, Mahlon Loomis, Nathan Stubblefield, James Clerk Maxwell and David Hughes. As in the case of the telephone, the invention of radio communication came about as a result of the work of a number of talented individuals and Guglielmo Marconi— like Alexander Graham Bell before him—had the luck to claim the credit. However, he did not have everything his own way. In 1943, the US Supreme Court invalidated one of Marconi's patents, claiming that Tesla and others had demonstrated "prior art". This decision is reminiscent of the resolution passed by Congress in September 2001 that officially recognized Antonio Meucci instead of Alexander Graham Bell as the inventor of the telephone.

The existence of radio waves had been suggested as early as the 1860s by the Scottish physicist, James Clerk Maxwell. By 1873, he had established the theoretical basis for the propagation of electromagnetic waves in a paper to the Royal Society. Starting from earlier work on electricity and magnetism by Michael Faraday and others, Maxwell produced a set of mathematical equations that described the behavior of electric and magnetic fields and their interaction with matter. Using these equations, Maxwell was able to show that oscillating electric and magnetic fields should be able to travel through empty space and that they would be reflected and refracted in the same way as light. Indeed, Maxwell calculated that the velocity of these fields through empty space would be very close to the known speed of light, leading him to speculate that light itself might be a form of electromagnetic wave. Since electromagnetic waves could oscillate at any frequency, he suggested that there might be a whole spectrum of electromagnetic radiation of which visible light was just one small part.

James Clerk Maxwell was born in Edinburgh in 1831 and graduated in mathematics from Trinity College, Cambridge, in 1854. During his life, he made significant contributions on the rings of the planet Saturn, the kinetic theory of gasses

James Clerk Maxwell.

(including the famous Maxwell's Demon[6]), optics and color photography. However, his most important work was his mathematical formulation of Michael Faraday's theories of electricity and magnetic lines of force and his establishment of the properties of electromagnetic waves at a time when nobody had any evidence to suggest that such waves even existed. His development of a quantitative link between electromagnetic phenomena and light is considered to be one of the greatest achievements of nineteenth-century physics. In 1931, on the 100th anniversary of his birth, Einstein described Maxwell's work as the "most profound and the most fruitful that physics has experienced since the time of Newton".[7]

Sadly, Maxwell died of abdominal cancer at the early age of 48, so he did not live long enough to see the experimental verification of his theory; it was still controversial at the time of his death and would remain so until the existence of electromagnetic waves was finally demonstrated in 1888 by a series of experiments conducted by Heinrich Hertz. Hertz was born in Hamburg in 1857 and he studied science and engineering at Berlin, Munich and Dresden,

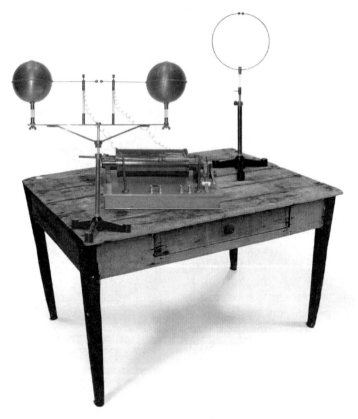

Recreation of Hertz's classic experiment. The transmitter is shown on the left, with the receiver on the right. The induction coil that drives the transmitter is in the center.

obtaining a PhD in 1880. His supervisor—Hermann von Helmholtz—was the inventor of the machine using oscillating tuning forks that had made such a deep impression on Alexander Graham Bell during his search for the "harmonic telegraph" (see Chapter 3). It was Helmholtz who suggested to Hertz that he might conduct some experiments to check the validity of Maxwell's predictions.

Hertz used a spark gap transmitter consisting of two polished metal balls with a small gap between them. When the two sides of the transmitter were connected to a high voltage generated by an induction coil, a spark was produced in the air gap. Two larger balls provided the transmitter with a degree of capacitance, which, when combined with the inductance of the coil, created a resonant circuit. As a result, each spark produced an oscillating electrical current in the apparatus and so a burst of electromagnetic waves was transmitted.[8]

The receiver used by Hertz was a piece of copper wire bent into a circle to form an aerial. A small brass sphere was attached to one end of the wire, while the other end of the wire was sharpened to a point. A screw mechanism was used to bring the point very close to the sphere without actually touching it, thereby creating a small air gap. If an electromagnetic wave was detected by the aerial, currents would flow in the wire and a spark would be seen at the air gap. As Hertz reported to the German Association for the Advancement of Natural Science in September 1889:

> "The sparks [at the receiver] are microscopically short, scarcely a hundredth of a millimetre long. They only last for about a millionth of a second. It almost seems absurd and impossible that they should be visible; but in a perfectly dark room they are visible to an eye which has been well rested in the dark."

Heinrich Hertz.

Hertz found that he could detect the electromagnetic waves over distances of up to 50 feet. In a series of brilliant experiments, he showed that the waves could be reflected, refracted and polarized just like light, and that they traveled at the same velocity as light. He also showed that interference patterns of maxima and minima could be produced, thereby allowing the measurement of wavelength. In this way, Hertz provided experimental confirmation that Maxwell's predictions had been correct and laid the groundwork for the development of radio as a practical method of communication.[9] Surprisingly, however, Hertz believed that his discoveries were of no practical significance and he did not live long enough to see the earth-shattering developments that they would precipitate. He died in 1894 at the early age of 36, possibly as a result of infectious substances left in his house by the medical clinic that had previously occupied it.

\* \* \*

In 1894, an Italian teenager called Guglielmo Marconi was reading an account of Hertz's experiments while on holiday in the Alps. Marconi had studied under Augusto Righi, a professor of Physics at the University of Bologna, who had conducted a number of experiments to investigate Hertzian waves. Unlike Hertz himself, Marconi was intrigued by the practical and commercial possibilities of the new discovery and he decided to conduct some experiments of his own. As he recalled many years later when he received his Nobel Prize:

> "I never studied physics or electrotechnics in the regular manner, although as a boy I was deeply interested in those subjects. ... At my home near Bologna, in Italy, I commenced early in 1895 to carry out tests and experiments with the object of determining whether it would be possible by means of Hertzian waves to transmit to a distance telegraphic signs and symbols without the aid of connecting wires. ... After a few preliminary experiments with Hertzian waves I became very soon convinced, that if these waves or similar waves could be reliably transmitted and received over considerable distances a new system of communication would become available possessing enormous advantages. ...[10]

The young Marconi established his first laboratory in the attic of his parents' country villa at Pontecchio, near Bologna. One evening in 1895, he summoned his mother up to the attic for a demonstration. He had set up a simple transmitter and receiver separated by a gap of about 9 meters. When he closed a switch at the transmitter, a bell rang at the receiver. He subsequently set up more powerful equipment in the gardens of the villa and was soon able to transmit messages over a distance of 2 kilometers. Since an intervening hill prevented any visual form of communication, successful reception had to be indicated by the firing of a gun.

At this point, it could be argued that Marconi was just one of a number of people who were investigating Hertzian waves. However, his energy, commercial acumen and financial resources soon meant that he pulled ahead of the field. Many of his competitors were only part-time radio experimenters (Oliver Lodge, for example, had a department to run at Liverpool University). Marconi, on the other hand, was dedicated to the task and would soon be able to establish his own company and employ technical staff to help him. Although his company ran at a loss for several years, Marconi could call upon financial backing from wealthy relatives and their business contacts.

Marconi's efforts to interest the Italian Department of Posts and Telegraphs in his discoveries were not successful—they already had a telegraph network and could see no need for radio. So, Marconi and his mother traveled to London. Marconi's mother was Annie Jameson, the Irish-born heiress to the Jameson whiskey fortune, who had settled in Italy and had married Guiseppe Marconi, the son of a wealthy Italian landowner. In England, she felt, her family might be able to exert more influence than in Italy.

Marconi was met in London by his cousin, Henry Jameson-Davis, who introduced him to A. A. Campbell-Swinton. After seeing a demonstration, Campbell-Swinton wrote a letter of introduction to W. H. (later Sir William) Preece, who was Engineer-in-Chief of the General Post Office. Preece allowed Marconi to establish a transmitter on the roof of the Central Telegraph Office in London, with a receiver on the roof of another building 300 yards away. On July 27, 1896, Marconi successfully demonstrated his system of wireless telegraphy. On September 2, the system was demonstrated again—this time over a distance of 1.75 miles on Salisbury Plain to representatives from the Post Office, the Army and the Navy.

In June of the same year, Marconi filed a British patent entitled "Improvements in Transmitting Electrical Impulses and Signals and in Apparatus Therefor". This is generally recognized as the world's first patent application for a radio telegraphy system, although it used various techniques developed by Nikola Tesla and it resembled an instrument developed by Alexander Popov. Indeed, Silvanus P. Thompson, an eminent electrical engineer of the time, later claimed that Marconi had never invented anything! However, Marconi did not claim to have discovered any new scientific principle; he had taken existing ideas, refined them and applied them in practical ways. As Marconi expressed it in the patent:

> "I believe that I am the first to discover and use any practical means for effective telegraphic transmission and intelligible reception of signals produced by artificially-formed Hertz oscillations."

The patent was issued in March 1897.

In the same year, Marconi established the first permanent wireless station at the Royal Needles Hotel on the Isle of Wight. He succeeded in communicating with two hired ferryboats and then with a station that he established at the Madeira Hotel in Bournemouth. The first paid wireless telegram was sent from the Isle of Wight to Bournemouth by Lord Kelvin[11] in early 1898. Lord Kelvin insisted in paying 1 shilling for the service, even though this technically constituted an infringement of the Post Office's monopoly.

In 1897, Marconi established the Wireless Telegraph and Signal Company Ltd with his cousin, Henry Jameson Davis. By the end of the following year, the company had opened the world's first wireless factory at Chelmsford in Essex. This company subsequently became Marconi's Wireless Telegraph Company Ltd, the forerunner of Marconi plc.

Marconi was keenly aware of the need to develop his discoveries into commercial applications, but he kept finding that his way was blocked by the claims of governments and the cable telegraph operators. When he discovered that there were no restrictions on communication with merchant shipping on the high seas, he began to develop wireless for use on board ships. In 1898, Marconi installed radio equipment on the Royal Yacht while the Prince of Wales (later Edward VII) was convalescing on board. This enabled Queen Victoria to communicate with her son from Osborne House on the Isle of Wight, and over 150 messages were exchanged during a period of 16 days. On less exalted ships, too, the radio was used mainly to allow passengers to send and receive messages, but the limited range and the small number of radio-equipped ships meant that it took some years for this business to become established. Marconi rented his system to ship owners on condition that only his personnel were allowed to operate the equipment and he refused to communicate with ships that were not using Marconi equipment. By 1913, when some nasty accidents had highlighted the safety advantages of radio at sea, Marconi had established an effective monopoly.

In March 1899, Marconi established the first radio telegraph link across the English Channel between South Foreland in Kent and Boulogne-sur-Mer in France. In the same month, the value of radio for managing emergencies at sea was illustrated when the East Goodwin Lightship used radio to summon the Ramsgate lifeboat to the assistance of the stranded German ship, *Elbe*, which had run aground on the Goodwin Sands in dense fog. A few weeks later, the lightship itself was rammed by the vessel *R.F. Matthews*, causing the international distress signal to be transmitted by radio for the first time.

British Patent No. 7777 (often referred to as the "four sevens patent") was awarded to Marconi's Telegraph Company Ltd on April 26, 1900. This patent described a method of preventing interference between adjacent radio stations and improving operating ranges by using two tuned circuits at the transmitter and two at the receiver[12]—a technique that turned out to be critically important for Marconi's subsequent attempts to communicate across the Atlantic. As before, the patent described improvements to the practical implementation of wireless telegraphy rather than any new scientific principle. It was the American version of this patent (US Patent 763,772) that was declared to be invalid by the Supreme Court decision in 1943. However, Marconi did not care—he had died 6 years earlier.

By 1901, Marconi had become obsessed with the idea of transmitting messages across the Atlantic, but there was a problem. As Marconi expressed it some years later:

> "At the time ... when communication was first established by means
> of radiotelegraphy between England and France, much discussion and
> speculation took place as to whether or not wireless telegraphy would
> be practicable for much longer distances than those covered, and a
> somewhat general opinion prevailed that the curvature of the Earth

would be an insurmountable obstacle to long distance transmission, in the same way as it was, and is, an obstacle to signalling over considerable distances by means of light flashes."[13]

This was a perfectly reasonable objection. Light beams cannot be used for communicating beyond the horizon because they travel in straight lines. Radio waves and visible light are both electromagnetic waves, so it was natural to assume that the same restriction would apply to radio waves. However, this line of argument failed to take account of the possibility of reflection. When you turn on an electric light in a room, some of the light will spread into adjacent rooms and corridors, even though there is no direct line between those places and the light bulb. This occurs because light is reflected off the walls and ceilings. Hertz had demonstrated that radio waves can also be reflected and it turns out that a layer in the Earth's atmosphere called the ionosphere can reflect radio waves at certain frequencies back towards the Earth. Under the right circumstances, very long-distance radio communication is possible using a series of reflections between the ionosphere and the Earth.

In January 1901, a successful test was carried out between St Catherine's Point on the Isle of Wight and The Lizard in Cornwall—a distance of 186 miles. The total height of each station was less than 100 meters above sea level, while the height required to give a clear line of sight between the two stations would have been more than 1,600 meters. Marconi therefore concluded that the curvature of the Earth would not prevent trans-Atlantic radio communication. He built a transmitter at Poldhu in Cornwall that was 100 times more powerful than anything previously constructed. The receiving station was located at St John's, Newfoundland, and it was there at about 12.30 p.m. on December 12, 1901, that he claimed to hear the first radio signal to be transmitted across the Atlantic:

> "Shortly before midday I placed the single earphone to my ear and started listening. The receiver on the table before me was very crude— a few coils and condensers and a coherer—no valves, no amplifiers, not even a crystal. But I was at last on the point of putting the correctness of all my beliefs to test. The answer came at 12: 30 when I heard, faintly but distinctly, pip-pip-pip. I handed the phone to Kemp: 'Can you hear anything?' I asked. 'Yes,' he said. 'The letter S.' He could hear it. I knew then that all my anticipations had been justified. The electric waves sent out into space from Poldhu had traversed the Atlantic—the distance, enormous as it seemed then, of 1,700 miles—unimpeded by the curvature of the earth."[14]

This claim is controversial, because modern understanding of radio propagation suggests that medium-wave signals could not have carried so far at that particular time of day. It is clear that Marconi heard something, but it was probably not the signal transmitted from Poldhu and it was some time before Marconi was able to communicate reliably across the Atlantic. On January 18, 1903, a radio link between Cape Cod, Massachusetts, and Poldhu was used for the first "official" radio communication between US President Theodore Roosevelt and King Edward VII. In 1907, a commercial radio telegraph service was opened between Glace Bay, Nova Scotia[15] and Clifden, Ireland. Three years later, communication was established between Clifden and Buenos Aires, Argentina—a distance of 6,000 miles. By 1918, the first message had been sent by radio from England to Australia.

The adulation that was showered upon Marconi has rather obscured the achievements of other radio pioneers and the Canadian Reginald Fessenden certainly deserves a mention at this point. While Marconi focused on radio telegraphy, Fessenden looked for ways to broadcast voice. Most of his contemporaries (including the great Thomas Edison) believed that this was impossible but in 1906, Fessenden proved them all wrong. At 9 p.m. on Christmas Eve, radio operators on ships in the Atlantic were amazed to hear a human voice coming from the equipment that they used to receive Morse Code. On that evening, Fessenden treated them to a recording of Handel's *Largo* played on an Ediphone, a rendering of *O Holy Night* played on his violin and a reading from the Bible. He then wished them a happy Christmas before bringing the first radio broadcast in history to a close. During the next 20 years, radio technology continued to develop rapidly, and radio-based telephone services started to appear. The first regular public trans-Atlantic telephone service opened on January 7, 1927, using long-wave radio and by 1930, radio telephone services were available from the United Kingdom to Australia, South Africa and Argentina.

In Chapter 1, we learned how Cooke and Wheatstone's telegraph played a central role in the capture of the murderer John Tawell as he traveled by train from Slough to London. In an analogous case, the murderer Dr Hawley Crippen and his mistress Ethel "Le Neve" Neave were arrested as they sailed across the Atlantic in July 1910. The captain of the *S.S. Montrose*, Henry Kendall, became suspicious of the couple and telegraphed the following message to his office in London using the Marconi radio system:

> "Have strong suspicions that Crippen—London cellar murderer and accomplice—are amongst saloon passengers. Moustache taken off. Growing beard. Accomplice dressed as boy. Voice manner and build undoubtedly a girl. Travelling as Mr and Master Robinson."

As a result of this message, Inspector Dew of Scotland Yard boarded the White Star liner *Laurentic* and managed to arrive at the St Lawrence river before the *S.S. Montrose*. Disguised as a tug boat pilot, he boarded the *Montrose* and arrested Crippen. Crippen was hanged at Pentonville Prison on November 28, 1910.

Two years later, wireless telegraphy played a critical role during the sinking of the *Titanic*. Two employees of the British Marconi Company (Jack Phillips and Harold Bride) were on the *Titanic* when it hit the iceberg and their distress calls were heard 58 miles away by the radio operator on the *Carpathia*. As a result, 705 passengers were saved who would otherwise have been lost. As Rt. Hon. Herbert Samuel, Postmaster General at the time, stated: "Those who had been saved, had been saved through one man, Mr Marconi and ... his marvellous invention."[16]

During the last 40 years of his life, Marconi was showered with honors. In 1909, he was awarded the Nobel Prize for Physics and in 1914 he became a member of the Italian Senate. In 1929, he received the hereditary title of Marchese. There were even a number of poems written about him, including one by Lord Dunsany (1878-1957), which included the following verse:

"Marconi came to birth
And before he left us again
Our voice was of greater worth,
For we speak and are heard plain
At the very ends of the earth."

Marconi continued to work right up to his death from a heart attack in 1937. During his funeral, wireless stations around the world fell silent for 2 minutes in tribute to the man

Guglielmo Marconi.

who had done so much to make global wireless communication a reality.

\* \* \*

By the middle of the twentieth century, a previously unexploited part of the electromagnetic spectrum was starting to provide many of the long-distance trunks required by telephone networks. Microwaves are electromagnetic waves with frequencies in the range 1–300 GHz (1 GHz is $10^9$ Hz). This means that they fit between radio waves and infrared in the spectrum, as shown in Figure 31.

The extremely high frequencies at which microwave links operate mean that they can carry large amounts of telephone traffic. The high frequencies also mean that the aerials required to transmit and receive the microwaves are relatively small.

As we learned earlier, some radio waves can follow the curvature of the Earth as a result of reflection off the ionosphere. This reflection does not work at microwave frequencies, so each microwave transmitter must have a clear line of sight to its receiver. For this reason, microwave towers are often built on high ground and are typically spaced about 30 miles apart.

In 1932, Marconi personally supervised the installation of the world's first microwave telephone link between Vatican City and the Pope's summer residence at Castel Gandolfo. The first commercial microwave link entered service in 1934 between England and France and covered a distance of 35 miles.

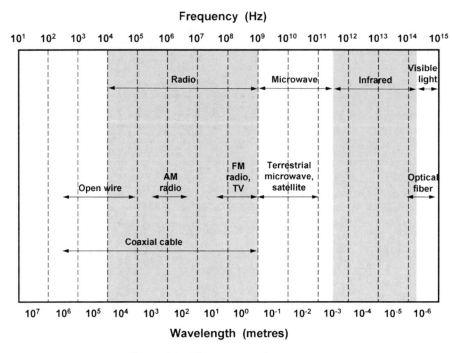

Figure 31. Electromagnetic spectrum.

By 1947, Boston and New York were linked by a multi-hop microwave system using seven towers and carrying 2,400 simultaneous telephone conversations.

The development of microwave occurred largely in parallel with the development of coaxial cable systems. However, microwave had lower construction and maintenance costs, so by the 1970s, microwave was carrying 70% of AT&T's long-distance voice traffic. Although microwave systems were subsequently displaced from core networks by the development of fiber optics, they are still used for high-capacity links over difficult terrain and microwaves continue to provide the connections between communication satellites and their associated ground stations.

\* \* \*

The concept of communicating via a satellite in Earth orbit was brought to public attention in 1945 by a letter published in *Wireless World* magazine. The author was an RAF electronics officer called Arthur C. Clarke, who later became famous for his science fiction classic *2001: A Space Odyssey*. Clarke proposed a global communications network based upon three satellites orbiting the equator at 35,786 kilometers above mean sea level—a height chosen so that the satellites would remain in a fixed position in the sky when viewed from the Earth. Satellites that have this property are now referred to as "geostationary". However, Clarke did not invent this idea—a Slovenian called Herman Potočnik

Microwave tower.

had written a number of articles on the subject as early as 1929. Furthermore, Clarke envisaged that the satellites would be large space stations manned by astronauts—a very different concept from the modern communications satellite!

In 1954, the idea of communicating via geostationary satellites was raised once again in a public lecture given by John R. Pierce[17] of AT&T, but the funding to develop the idea was not available at the time. It was the Soviet Union's launch of Sputnik in October 1957 that caused attitudes in the United States to change and by August 1960, an experimental communication satellite had been launched. Echo was a 100-feet diameter balloon with an aluminum coating that could reflect microwaves. Since it was placed in a low orbit, it did not remain stationary in the sky and systems were required to capture and track the fast-moving satellite every time it appeared above the horizon. To demonstrate the operation of the system, a pre-recorded message from President Eisenhower was transmitted coast to coast from Goldstone, California, to Crawford's Hill, New Jersey.

However, before Echo had even been launched, Pierce had already turned his attention from "passive" satellites like Echo to "active" communications satellites. Passive satellites simply reflect any received signals, while active satellites amplify the received signals before re-transmitting them on a different frequency. Active satellites are more complex to build, but give much better performance. President Kennedy's speech in May 1961 famously committed the United States to put a man on the Moon before the end of the decade, but it also committed the country to build a global satellite communications system and it added $50m to NASA's budget for this purpose. An active communications satellite, Telstar I, was launched into orbit in July 1962. As in the case of Echo, Telstar I was not in a geostationary orbit and could support trans-Atlantic communications for less than 2 hours per day. A constellation of about 55 satellites would have been needed to provide 99.9% coverage between any two points on Earth, but only one of these (Telstar II) was ever launched.

Telstar demonstrated the potential of communications satellites. It could support up to 480 simultaneous telephone conversations and on the day after it was launched, it was used for the first trans-Atlantic television transmission. Ground stations with large dish aerials were developed to support satellite communications, including the British Post Office's Goonhilly Downs facility in Cornwall.

The International Telecommunications Satellite Organisation (INTELSAT) was established in 1964 to develop global satellite communications and by 1999, more than 100 countries had become members of the organization. In April 1965, Intelsat1 (Early Bird) was launched. It was the world's first commercial geostationary communications satellite and carried roughly 10 times the capacity of a submarine cable at about one-tenth of the cost. This price advantage was maintained by subsequent Intelsat satellites until 1988, when the first fiber optic cable was laid across the Atlantic.

Although satellites were cost-effective, they had one significant weakness: transmission delay. An orbit of 35,786 kilometers above mean sea level is required to keep a geostationary satellite in a fixed position above the Earth's surface. Traveling at the speed of light, microwaves take roughly a quarter of a second to make the journey from Earth to the satellite and back again. This delay can cause telephone conversations to become stilted and difficult, and the problem was a familiar feature of trans-Atlantic telephony during the 1980s. The problem is much less common today because satellites have largely been replaced by fiber optic cables on telephone trunk routes. However, satellite transmission continues to be used to provide telecoms services in remote or inaccessible areas.

*  *  *

On June 3, 1880, Alexander Graham Bell transmitted the first wireless telephone message using his newly invented "Photophone". The Photophone contained a flexible mirror that vibrated in response to the speaker's voice. When sunlight was projected onto the mirror, the vibrations were superimposed upon the reflected beam. At the receiver, crystalline selenium cells were used to detect the light beam. The resistance of these cells decreased as the intensity of the light increased, thereby allowing the speech signal to modulate an electrical current. This current could then be converted back into audible speech using an ordinary telephone receiver. In effect, the Photophone was a telephone that used light instead of electricity to carry the signal from the transmitter to the receiver.

Bell took out four separate patents relating to the Photophone and it is reported that he considered it to be a more important invention than the telephone because it could operate without wires. However, his design was vulnerable to interference from rain or fog and it could only communicate over a restricted distance. It was not until the development of fiber optics that communication using a beam of light became a practical possibility.

As far back as the 1840s, it had been found that light could be guided along inside a jet of water and this discovery had been used to illuminate elaborate public fountains. By the end of the nineteenth century, dental illuminators had been developed that used quartz rods to carry light around corners. During the 1920s, John Logie Baird in England and Clarence W. Hansell in the United States had patented the idea of using arrays of transparent rods to transmit television or facsimile images. Attempts to develop a flexible fiber optic gastroscope for viewing inside a patient's stomach had begun in the early 1930s—rigid gastroscopes having been described as "one of the most lethal instruments in the surgeon's kit"—but the first semi-flexible gastroscope was not patented until 1956. This was soon followed by a range of medical and non-medical imaging applications for optical fibers, such as inspecting the welds inside reactor vessels.

By the early 1960s, telecommunications engineers were once again looking for ways to increase the bandwidth on trunk routes. Coaxial cable and microwave links were heavily loaded and demand was expected to grow as networks carried increasing amounts of telephone and television traffic. Higher-frequency microwaves would offer higher capacity[18] but they also suffered from increased attenuation caused by the atmosphere, so researchers began looking at the possibility of confining these microwaves within hollow pipes that would act as waveguides. It was recognized that transmission systems based upon visible light would have a much higher potential capacity than systems based upon microwaves, but light signals transmitted through the air were attenuated by rain, snow and fog, and fiber optic cables introduced far too much attenuation to permit communication over long distances.

There was, however, a group of researchers at STC's laboratories in Harlow, Essex,[19] who were not prepared to accept the negative view of fiber optics that prevailed in the telecommunications industry. Charles Kao and George Hockham were the first to recognize that the attenuation of contemporary fibers was caused by impurities rather than by any intrinsic limits in the glass itself. In a famous paper published in July 1966,[20] they suggested that if the level of impurities could be reduced sufficiently, then optical fiber might have "important potential as a new form of communication medium". At the time, many people were openly skeptical, but the low level of attenuation required was achieved 4 years later by researchers at Corning Glass Works in the United States.[21] This triggered the development of fiber optics for telecommunications applications and by 1978, a fiber optic system had been installed between the Post Office Research Centre at Martlesham Heath and the nearby telephone exchange at Ipswich. In 1984, the world's first 140 Mbit/second optical fiber system was operating between Luton and Milton Keynes, and the Isle of Wight was linked to the mainland by a submarine fiber optic cable in the following year. The first trans-Atlantic submarine cable to use fiber optics (TAT-8) went into operation in 1988. In 2009, Charles Kao was awarded the Nobel Prize for Physics "for groundbreaking achievements concerning the transmission of light in fibers for optical communication".[22]

Optical fibers are glass[23] rods that have been stretched until they are thinner than a human hair. Hundreds of fibers can be grouped together into a single cable that still retains sufficient flexibility to bend around corners. Fibers are generally used in pairs (one fiber operating in each direction), but it is possible for a single fiber to support bi-directional communication. The light carried by a fiber optic cable is generated either by a semiconductor laser or by a light-emitting diode (LED).[24] At the receiver, the light is converted back to an electrical signal using a light-sensitive device such as a photocell or photo-diode—in a way that is reminiscent of Alexander Graham Bell's Photophone.

If a beam of light is introduced at one end of the fiber, the light will be trapped inside the fiber by a process called total internal reflection.[25] Figure 32 shows an optical fiber guiding light around a bend, illustrating that fiber optic systems can be used in situations in which a direct line of sight is not available.

Notice that Path A in Figure 32 is significantly shorter than Path B. This means that if a pulse of light is transmitted simultaneously on Path A and Path B, then the Path A component of the pulse will arrive at the receiver before the Path B component. As a result, the pulse that emerges at the receiver will contain a contribution from Path A plus a delayed contribution from Path B. It will also

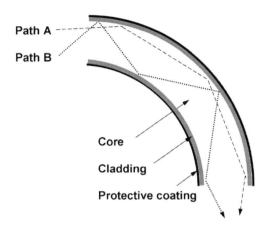

Path A

Path B

Core

Cladding

Protective coating

**Figure 32.** Paths of light inside an optical fiber.

contain contributions from all the other available optical paths, each delayed by a different amount. This distortion of the pulse shape is known as "dispersion".

Dispersion is a serious problem, because it limits the speed at which information can be communicated down a fiber. It is a bit like trying to communicate by shouting down a long valley. The person at the far end will hear the speech that has traveled directly down the valley, but the message may become unintelligible if large numbers of echoes arrive after bouncing off the sides of the valley. The best way to overcome this problem is to shout one word at a time, and to wait long enough between each word for the echoes to die away. It is a similar situation with optical fibers—each bit of data must be transmitted for long enough to ensure that the receiver will receive it correctly. Although optical fibers can carry huge numbers of bits per second, dispersion places upper limits on the speed of operation and/or the maximum length of the fiber.[26]

Fiber that supports multiple optical paths is referred to as "multimode". This type of fiber has the advantage of being cheap, but the dispersion caused by multiple path lengths restricts its use to distances less than about 500 meters. However, if the core of an optical fiber is reduced in diameter, the point is eventually reached at which light is restricted to a single path down the fiber. The principal advantage of this "single mode" fiber is that the dispersion caused by multiple path lengths is effectively eliminated,[27] thereby making it suitable for use over the much longer distances required by telecoms networks. Inevitably, these advantages come at a price. Since the diameter of the core is now measured in microns (a micron is a millionth of a meter), it becomes much more difficult to achieve an effective optical coupling between the fiber and the laser that drives it. This pushes up the cost of working with single mode fiber.

The problems of dispersion have been further reduced by the introduction of solitons into optical systems. A soliton is a wave with a particular shape that allows it to propagate through certain media without experiencing dispersion. The first recorded observation of a soliton was made by John Scott Russell in 1834 on a canal in Scotland. He was watching a boat being drawn along by a pair of horses when, for some reason, it suddenly stopped. Scott Russell noticed that the bow wave continued to move forward "at great velocity, assuming the form of a large solitary elevation, a well-defined heap of water which continued its course along the channel apparently without change of form or diminution of speed". He followed the wave on horseback for more than a mile, convinced that he had made an important new discovery. However, contemporary science showed little interest and it was not until the 1960s that the significance of the soliton was finally recognized.

The solitons used in optical communications are narrow, high-powered pulses of light. Since they retain their shape over long distances, solitons can travel much further before electrical regeneration becomes necessary. However, they are still vulnerable to the normal effects of attenuation as they travel down the fiber, so amplification is required at regular intervals to boost them up again. Fortunately, it is possible to provide this amplification without the need to convert the signal from optical to electrical and back again. The device that achieves this impressive feat is called an optical amplifier.

Optical amplifiers have much in common with lasers. In both cases, a material is "pumped" so that incoming photons with a specific wavelength will stimulate the emission of further photons at the same wavelength. An optical amplifier consists of a length of fiber that has been "doped" with a small amount of impurities and amplification is achieved by stimulating the emission of photons from the dopant ions. This is done by pumping the doped fiber with a laser to excite the electrons into a higher energy band, from where they can return to their former energy band by emitting a photon. Every photon arriving at the input of the optical amplifier triggers a much larger number of photons at the output, thereby achieving signal amplification. Although many different types of optical amplifier have been developed, those used for telecommunications applications typically use a rare earth called erbium as the dopant because it produces light at frequencies that suit the characteristics of optical fibers. The first erbium-doped fiber amplifier was invented at Southampton University in 1986.

Many of the high-speed fiber links deployed during the mid 1990s operated at 2.4 Gbit/second (i.e. 2,400,000,000 bit/second). Although this sounds like a very impressive number, it still represents only a tiny fraction of the capacity of an optical fiber. In Chapter 4, we learned how Time Division Multiplexing and Frequency Division Multiplexing have been used to increase the capacity of a copper cable. In the case of optical fibers, both techniques can be used simultaneously! Time Division Multiplexing is used to combine a large number of telephone channels (or data, or television channels, or whatever) into a high-speed signal running at about 10 Gbit/second. A technique called Dense Wavelength Division Multiplexing (DWDM)—an optical form of Frequency Division Multiplexing—is then used to transmit large numbers of these 10 Gbit/second signals down a single fiber by using a different wavelength of light for each signal. With this approach, the capacity of a fiber can be extended into the terabit range (1,000,000,000,000 bit/second).

The astonishing success of fiber optic technology was a key factor behind the Internet bubble in the late 1990s. Since its earliest days, the telecommunications industry had worked on the assumption that bandwidth was scarce and expensive. Suddenly, thanks to fiber optics, bandwidth became plentiful and cheap. This triggered the development of many new high-bandwidth applications that had previously been impossible. However, the speed with which DWDM technology developed and the extraordinary rate at which new fiber was installed meant that the supply of bandwidth grew faster than the demand created by the new applications. As a result, prices collapsed and many network operators went out of business. Fortunately, the fiber that they installed has not gone away—it is still buried in the ground, waiting for the day when it will be needed. Some of the key advantages of fiber optics are summarized in Table 5.

It is easy to see from this table why fiber has almost completely displaced copper cables in core networks and is spreading rapidly into regional and access networks. Fiber has also displaced microwave systems on long-haul routes—except over difficult terrain where it is impractical to install cables.

**Table 5.** Advantages of fiber optics.

| | |
|---|---|
| Capacity | An optical fiber can provide many thousands of times more capacity than a copper cable. |
| Attenuation | The attenuation of an optical fiber can be 200 times lower than the attenuation of an equivalent length of coaxial cable. Unlike microwaves, the attenuation of an optical fiber is not affected by the weather. |
| Dispersion | Electrical regeneration is often unnecessary on long-haul fiber routes. This saves cost and means that a link can often be upgraded by just changing the equipment at each end—the fiber and the optical amplifiers remain the same. |
| Crosstalk | Electrical signals traveling down a copper wire can induce currents in other wires in the same cable. The pulses of light traveling down a fiber cause no interference on other fibers in the cable. |
| Electrical interference | Copper cables can pick up interference from adjacent power cables. This can interfere with communication and may damage electronic equipment at either end of the cable. Optical fibers have high electrical resistance and so can be used safely in strong magnetic fields. |
| Safety | Optical fibers do not carry electricity, so they can be used safely in hazardous areas near flammable liquids or gasses. |
| Security | An optical fiber produces no electromagnetic radiation and it is very difficult to tap an optical fiber without disrupting the operation of the link. This means that fibers are ideal for links requiring high security. |
| Life expectancy | A coaxial cable typically lasts for about 10 years before it has to be replaced. An optical fiber can remain in service for far longer. |
| Size | Optical fibers are thinner than copper wire. Up to 1,000 fibers can be bundled together into a single cable. |
| Weight | An optical cable weighs less than a comparable copper cable. |
| Cost | Optical fiber can be manufactured more cheaply than an equivalent length of copper wire. However, some of the operations required to install a fiber cable (e.g. splicing) are more expensive than the equivalent operations for copper cables. |

However, it is still relatively rare to find fiber installed all the way to the customer's home or office. The reasons for this will be explored in the next chapter.

# 7 The last mile

As we saw in Chapter 4, the primary role of a telephone access network is to provide the connections between subscribers and their local exchange. In telephone parlance, this part of the network is known as "the last mile". As early as 1878—a mere 2 years after the invention of the telephone—Alexander Graham Bell was already speculating on the need for an access network:

> "It is possible to connect every man's house, office or factory with a central station, so as to give him direct communication with his neighbours .... It is conceivable that cables of telephone wires could be laid underground, or suspended overhead, connecting by branch wires with private dwellings, shops, etc., and uniting them through the main cable with a central office[1]."[2]

To a modern telephone engineer, these words read like a statement of the obvious, but they must have seemed wildly fanciful when they were written. At that time, most telephones were still connected in pairs to provide nothing more than an intercom between two buildings, so the distinction between access and core networks did not really exist. Furthermore, the telegraph network—which was well developed by 1878—did not have an access network. Extending telegraph wires to people's homes would have been pointless, because most of them were not skilled in Morse Code. Instead, telegrams were printed at the local telegraph office and were delivered to their recipients by messenger boys. The telephone was different because it required no particular skill to use—and it was impractical to expect people to turn up at their local telephone exchange just to receive a phone call!

We learned in the previous chapter how copper wire rapidly displaced galvanized iron and steel wire in core networks. Not surprisingly, the same thing also happened in access networks. However, while the core network evolved through a succession of new technologies—such as coaxial cable, microwave and optical fiber—the access network continues to this day to be dominated by copper wire. This indicates that the access network has some very different characteristics from the core. Copper wire may not be able to support the gigabits of traffic found on trunk routes, but it is quite adequate to carry a single telephone call between a subscriber and the local exchange. Unlike optical fiber, copper wire can also carry electrical power, which is why telephones continue to work during a power cut. Making joints in copper wire is much simpler than splicing optical fibers and the equipment required at each end of the line is significantly cheaper. Finally, the majority of buildings in developed countries

A. Wheen, *From Dot-Dash to Dot.com: How Modern Telecommunications Evolved from the Telegraph to the Internet*, Springer Praxis Books, DOI 10.1007/978-1-4419-6760-2_8, © Springer Science+Business Media, LLC 2011

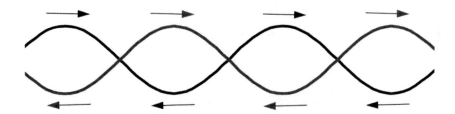

**Figure 33.**  Canceling induced currents in twisted pair cable.

already have copper telephone wires and the telephone companies are not anxious to write off this massive investment.[3] There are, of course, some situations in which more exotic access technologies are necessary—and we will come to those later in this chapter—but copper wire is still perfectly adequate for delivering services to most domestic customers and many small businesses. The launch of multi-megabit broadband services suggests that copper wire still has some life left in it.

The type of copper wire used in the access network is referred to as "twisted pair" and was originally patented by Alexander Graham Bell in 1881. In modern twisted pair cables, each wire is insulated by a plastic coating and pairs of wires are then twisted together to make them less vulnerable to electrical interference. The principle is illustrated in Figure 33.

If the twisted pair finds itself in an alternating magnetic field (from a nearby power cable, perhaps), then the induced currents at a particular moment in time might appear as shown by the arrows in the diagram. However, if you look at the currents induced in just one of the wires, you will see that they are going in opposite directions and so will tend to cancel each other out. This neat trick provides twisted pair cable with a degree of immunity from interference.

Although twisted pair cables are traditionally used for carrying analog telephone calls, they are also capable of carrying digital services such as broadband. Using a technology called ADSL,[4] a single twisted pair cable can support an analog telephone line AND broadband Internet access at the same time. It achieves this impressive feat by transmitting the broadband data at frequencies above those required for the voice channel.[5] As explained in Appendix E, ADSL is just one member of a family of technologies that can deliver digital services over ordinary telephone lines.

The telephone line coming into your home probably contains two twisted pairs, of which only one pair is actually used. A typical arrangement for connecting BT customers back to their local exchange is illustrated in Figure 34.

The telephone lines from a group of homes are brought together at a distribution point (DP), which is often mounted at the top of a telegraph pole.[6] From the DP, a multi-pair cable connects the homes to a primary cross-connection point (PCP), which usually takes the form of a street cabinet. Cables from a large number of homes are brought together at the PCP and are connected back to the local exchange by a cable containing hundreds of wires.

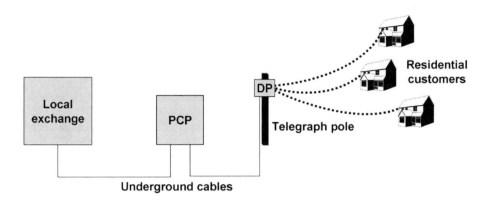

**Figure 34.** BT access network.

The huge cost of building an access network based on copper wire explains why very few network operators have ever attempted to do it. Most western European countries developed their copper access networks at a time at which telecoms was a state-owned monopoly, so the costs were borne by the long-suffering tax payer. When telecoms markets were opened up to competition in the 1980s and 1990s, access networks became the property of privatized former monopolies such as BT, France Telecom and Deutsche Telekom. This created rather lop-sided markets, because the company that owned the access network was inevitably more equal than its competitors. Governments and regulators looked for ways to address this imbalance by making the copper access network available for use by competing network operators. Some of their more successful interventions in the market are described in Appendix F.

In spite of the prevailing view that the access network was a natural monopoly, an attempt was made during the 1980s to create a competing access network in the United Kingdom by allowing cable TV networks to carry telephone services. It was argued that two separate revenue streams (telephony and cable television) would make it possible to justify the costs involved. The way in which the cable TV operators set about building their access networks departed significantly from traditional designs, as can be seen in Figure 35.

While traditional access networks use twisted pair copper cables all the way from the local exchange to the subscriber, cable TV networks use optical fiber from the cable TV headend as far as a street cabinet. A "siamese cable" containing coaxial cable (to carry the television and broadband[7] signals) and twisted pair (to carry the telephony) is used to link the street cabinet to each home in the vicinity. A multiplexer in the street cabinet converts the optical signal arriving from the headend into electrical signals for delivery to each home. Street cabinets typically serve between 60 and 480 homes and have become a familiar sight in cable TV franchise areas.

The architecture adopted by cable TV networks has a number of advantages. To begin with, it is inherently more reliable than traditional access architectures

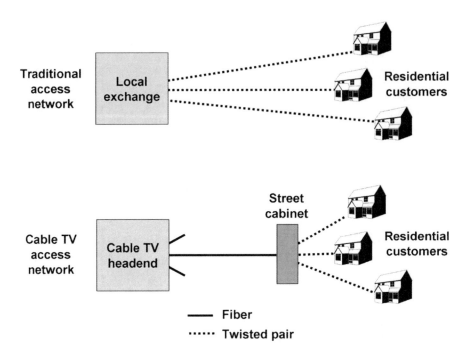

**Figure 35.** Comparison of cable TV and traditional access network.

and the electronics housed in the street cabinet can help to identify any faults that do occur. Second, the copper cables are much shorter than in traditional access networks, so they have the potential to deliver higher-speed digital services.[8] Finally, there is no real restriction on the area that can be served by a single cable TV access network, because the street cabinets are connected back to the network using fiber rather than copper. All customers on a cable TV access network can receive full-speed broadband services—irrespective of how far they live from the cable TV headend. As a result of these advantages, a number of traditional access networks are migrating towards cable TV's "Fiber to the Cabinet" architecture.

\* \* \*

With optical fiber turbo charging the core network, the copper-based parts of the access network have become an increasingly serious bandwidth bottleneck. The cable TV architecture extends optical fiber into the access network as far as the street cabinet and even higher bandwidths could be achieved if optical fiber was extended all the way to each home. Unfortunately, this is much easier said than done. Whilst it is cost-effective to install optical transmitters and receivers on routes carrying thousands of simultaneous telephone calls and huge amounts of Internet traffic, the costs are much harder to justify at the end of

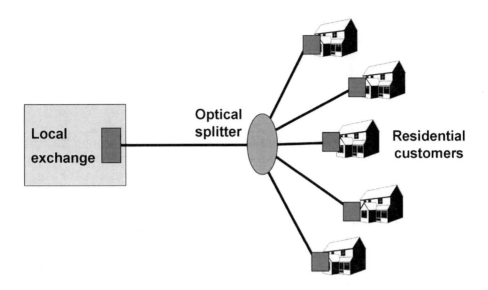

**Figure 36.** Passive optical network.

a telephone line serving a single household that may only make a few phone calls per day.[9]

Passive Optical Networks (affectionately known as PONs) provide a possible solution to this problem. The basic concept is illustrated in Figure 36.

A single fiber carries bi-directional traffic between the local exchange and an optical splitter located in the street. The splitter is simply a piece of glass, so it does not require a power supply and does not need to be housed in a street cabinet. Transmissions from the local exchange are received by all the homes on the PON, but each receiver will reject messages that are not addressed to it. In the upstream direction, each transmitter is given a separate opportunity to transmit in order to prevent interference. A PON might typically serve 32 homes and cost savings are possible because it is very much cheaper to provide one optical transmitter and receiver at the local exchange instead of 32.

In the early 1990s, several large network operators saw PONs as a way of bridging the wide cost difference between copper and fiber. Network operators were keen to gain experience of using fiber in the access network and it was expected that PONs would improve network reliability. Unfortunately, these hopes were not realized. This was partly because the true costs of the technology turned out to be higher than expected and the technical and operational issues were more challenging. There was also some resistance from customers, because PONs delivered no additional services and no price advantage. In areas in which the copper network was already well established, PONs struggled to compete.

In recent years, PONs and other "Fiber to the Home" technologies have started to find a role in up-market housing developments. The burgeoning demand for high-definition television services and very high-bandwidth connections to the Internet suggest that copper wire may eventually be displaced by fiber. However, don't expect change to happen overnight—copper wire has been in telephone networks since 1877 and it will be many years before it finally disappears.

\* \* \*

A very significant factor in the cost of building an access network is the cost of burying all that cable in the ground—or suspending it from telegraph poles. A radio-based access network seems like an obvious solution to this problem. We are all familiar with the use of radio in mobile phone networks, but radio in the fixed access network is much less common. For many years, "fixed wireless access" has been hyped as "the next big thing". Has its time finally come?

In 1991, a company called Ionica was established in Cambridge, United Kingdom. The objective of the new company was to build a fixed wireless access network to deliver telephone services in direct competition with BT. Although Ionica's prices were slightly lower than BT's, they aimed to attract customers primarily by offering better service features. For example, Ionica customers received a free second line so that two people in the household could make phone calls at the same time. Customers were also offered a separate telephone number with a different ring tone for each member of the household—so that they could tell from the sound of the ringing who should answer the phone.

The distinctive aerial used to receive the Ionica service.

These features must have been a blessing for families with teenage children and would also have been useful for many small businesses. The financial community loved the concept and were desperate to invest. Ionica floated on the stock exchange in July 1997 and by the peak of the dot.com boom, its value had soared to over a billion pounds. However, in common with many start-up technology companies of that period, Ionica ran into financial difficulties. The company went into administration in October 1998 and service was finally terminated to the last remaining customers in February 1999.[10]

The lesson from this sorry tale is that it is extremely difficult to sustain a new access network in places where an existing access network is already well established. Fixed wireless access is an option for delivering telephone services in rural areas where copper cables would be excessively long and Ionica's radio technology was in fact used successfully in remote parts of Canada before it was brought to the United Kingdom. Wireless access is also an option in developing countries where the copper network is limited or non-existent, but many of these countries are simply building mobile networks instead. Fixed wireless access can occasionally find applications in greenfield housing projects where no access network currently exists, but these developments tend to favor more exotic fiber-based technologies. If you want to deliver high-bandwidth services to large numbers of homes in an urban area, then fixed wireless is probably not the right way to go—apart from anything else, radio spectrum is a scarce resource, and it is better to keep it for applications that require mobility. Having said that, radio does have some applications in the fixed access network and these are discussed in Appendix G.

* * *

Even if future developments lead to substantial improvements in the access network, there will still be many inaccessible areas where wireless broadband coverage is not available and where wired broadband services are not feasible. For people living on remote Scottish islands, broadband delivered by satellite may be the only available option. Satellite-based broadband services have traditionally suffered from the need to use a dial-up return path, but advances in technology now allow true two-way broadband services to be offered over satellite. Satellite broadband services have also been more expensive than other forms of broadband, but this was because the satellites were designed to provide coverage over a wide geographic area—which is exactly what you want for TV broadcasting. Satellites designed for Internet access need much smaller coverage areas so that each frequency can be re-used in different locations, thereby increasing the overall capacity of the system. Satellites of this type are now starting to appear and could make satellite broadband an increasingly cost-effective proposition for people in rural locations.

In the future, unmanned solar-powered airships called High Altitude Platforms (HAPs) may possibly provide an alternative method of delivering broadband services to remote communities. Whilst geostationary satellites orbit 35,786 kilometers above the Earth, HAPs would be permanently located at an altitude of about 20 kilometers—thereby avoiding the long transmission delays associated with satellites. As in the case of some satellites, the coverage from a HAP would be restricted to a small area so that the same radio frequency can be re-used many times in different parts of the country.

HAPs would communicate with the ground using microwave radio, as satellites do. However, HAPs would also be required to communicate with each other, and they would do this using beams of light. Well up in the stratosphere, the light beams would experience no interference from fog, snow, rain, passing birds or commercial aircraft, and the curvature of the Earth would be less of a restriction than it is on the ground.[11]

HAPs are still at the research stage. The European Union has funded research into HAPs as a means of delivering broadband connections to rural parts of Europe, with the longer-term goal of delivering high-bandwidth connections to moving vehicles. HAPs could also be used to provide temporary broadband coverage for a major event or at the site of a natural disaster. It is claimed that HAPs are cheaper and more efficient than other forms of broadband because they do not require underground cabling or masts, but their commercial viability remains to be proven.

Powerline communication is another access technology with an outside chance of success. It uses the electricity distribution network for delivering voice and data services into people's homes. If a low-frequency signal is injected at an electricity substation, it will propagate to all parts of the local distribution network and can be used (for example) to switch street lamps on and off. The frequencies used have to be low (less than 1 kHz) in order to pass through the electricity transformers and this low frequency places severe restrictions on the bit rate that can be supported. For voice or data services, a much higher carrier frequency is required, so the signal has to be injected downstream of the last transformer that serves a group of homes. This is, in effect, a radio system using power cables as the carrier medium.

Between 1994 and 1996, Nortel Networks and Norweb conducted a trial of powerline communications for delivering voice and data services to 20 homes in the Greater Manchester area. To ensure reliable operation, "conditioning units" had to be installed in every home connected to the network (not just those homes where service was actually required).[12] The high cost of providing the conditioning units led to the decision not to proceed with the launch of a commercial service, despite the technical success of the trial. However, conditioning units are far less essential for broadband services than they are for voice, because the protocol will re-transmit any data that gets corrupted by noise. By 1998, the rapid growth in demand for high-speed Internet access, combined with the prospect of reduced conditioning costs, convinced Nortel Networks and Norweb to establish a joint venture called Nor.Web.

Nor.Web soon ran into technical difficulties. It appears that one of the problems concerned the size and shape of British street lamps, which made them ideal for broadcasting powerline signals. This caused serious interference on some BBC services and also affected amateur radio and emergency bands. In the light of these difficulties, the Institution of Electrical Engineers strongly opposed the introduction of powerline communications in the United Kingdom and Nor.Web finally closed in September 1999. It was claimed that the technology had been shown to work but that projected volumes and profitability were insufficient to justify the investment required. Since then, other trials have taken place in the United Kingdom, but no significant commercial broadband services have been established.

Powerline technology has the advantage that it can be delivered to any power socket in the house, thereby eliminating the need for separate in-house cabling.[13] It is also claimed that the problem of radio interference can be cured by filtering out those frequencies that cause problems. There have been modest deployments in a number of countries and powerline could develop into a significant competitor for ADSL and cable broadband if the technical, regulatory and commercial issues really can be resolved. However, historical precedents suggest that a considerable degree of skepticism may be justified.

# 8  Computers get chatty

The telegraph was the first true data network. Although it was normally restricted to just one form of data (text), it was certainly capable of carrying others (such as facsimile). It could not, however, carry voice. It was not until the arrival of the telephone in the late nineteenth century that voice started to challenge data as the dominant form of traffic on public networks. By the middle of the twentieth century, networks were overwhelmingly designed to carry voice, but within 50 years, the pendulum was swinging back towards data. This occurred because developments in digital computing and the arrival of the Internet meant that computers were now able to talk to other computers. They turned out to have a lot to say! Today, data is once again the dominant form of traffic on most networks. In this chapter, we introduce some of the key features of data networks.

Let's begin by addressing one of the most fundamental and far-reaching differences between voice and data networks: the distinction between circuit switching and packet switching. Traditional telephone networks operate by setting up a circuit between two telephones to carry a telephone call. In the early years of the telephone network, circuits were set up by human operators using switchboards and patch chords to make physical connections. More recently, digital circuits have been set up and cleared down using electronic switches under software control. Irrespective of whether they are set up electronically or manually, the fundamental principle remains the same: from the moment they are established to the moment they are cleared down again, circuits give the user a fixed amount of dedicated bandwidth. In digital networks, this means that the user is provided with a steady flow of bits at a fixed bit rate. These bits will continue to flow while the circuit is established—irrespective of whether they are carrying any useful information or not.

Packet switching networks are entirely different. Rather than transmitting a continuous stream of bits, packet switching networks transmit their data in discrete lumps known as packets. A packet of data is rather like a physical packet that is posted to a remote destination. The physical packet carries the source and destination address, along with information about its level of priority (first/second-class mail) and how it should be handled (e.g. fragile). Physical packets can be large or small, but there are limits on the sizes that the postal system will accept. In some cases, a large item might have to be broken down and posted in several separate packets, but it could then be re-assembled when all the packets have arrived at their destination. For valuable items, confirmation would be required that each packet has been safely delivered.

A. Wheen, *From Dot-Dash to Dot.com: How Modern Telecommunications Evolved from the Telegraph to the Internet*, Springer Praxis Books, DOI 10.1007/978-1-4419-6760-2_9,
© Springer Science+Business Media, LLC 2011

All of these characteristics of the postal system (and many others) have exact equivalents in packet switching networks. Each packet of data carries a header showing its source and destination address, along with its level of priority and other important details. The rest of the packet can carry any form of digital information (voice, data, text, video, etc.), just as a physical packet can contain almost any type of small object. Continuous streams of information (such as voice conversations) are broken down into conveniently sized lumps for transmission in separate packets, with the lumps being re-assembled at the destination. If required, transmission protocols can provide the sender with confirmation that each packet has arrived safely at its destination.

In the postal system, sorting offices linked by high-speed road or rail services are used to ensure that each packet moves rapidly from its source to its destination. In packet switching networks, the road or rail links are replaced by transmission links, while the sorting offices are replaced by devices called "routers". To illustrate how this works, let's consider how a packet makes its way from Computer A to Computer B across the packet switching network shown in Figure 37.

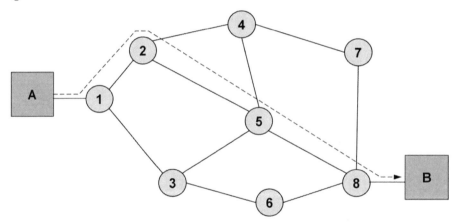

**Figure 37.**   Route across packet switching network.

When the router at Node 1 receives the packet from Computer A, it analyzes the header to decide how best to route it towards its destination. In this particular case, there appear to be three good routes across the network (1-2-5-8, 1-3-5-8 and 1-3-6-8). There are also several other routes (such as 1-2-4-7-8) that appear to be less good because they pass through a larger number of nodes. Although the router will consider the number of nodes when making its decision, it may also take other factors into account such as the available bandwidth on each route, the level of traffic congestion, the transmission delay and the bit error rate. It will also avoid any route that it knows is faulty. Since these conditions can change over time, it is quite possible for two packets with the same source and destination to follow different routes across the network. In some cases, this can result in packets arriving out of sequence, so the receiver must be able to sort them out.

Although the route selection process can be relatively sophisticated, the router at Node 1 only has two possible choices: it can send the packet to Node 2, or it can send it to Node 3. In the example illustrated in Figure 37, it decides to transmit the packet to Node 2. The router at Node 2 will then examine the packet header before deciding how to forward it on the next leg of its journey. Once again, there are two possible options and the router decides to send the packet to Node 5. The router at Node 5 has three possible options, but one of them looks much better than the other two for a packet that is heading towards Computer B. The packet is therefore forwarded to Node 8.

Packet networks can operate at very high speeds and networks linking many millions of end points can be constructed. Furthermore, packet networks can handle bursty traffic efficiently—if there is no data to carry, then no packets will be sent and the network resources will remain available for other users. We can illustrate this by considering a specific example of a packet network: the Internet. If you request a web page while surfing the World Wide Web, let's suppose that 100 Kbytes of data need to be transferred to your PC. You expect the page to appear within a couple of seconds so—to rather over-simplify the situation—the bandwidth required is about 50 Kbyte/second. Since there are eight bits in a byte, this equates to 400 Kbit/second. However, you might then go off to lunch and not request another web page for several hours. If the network permanently allocated a 400 Kbit/second circuit to each user, the whole thing would be hopelessly inefficient. Instead, data is sent across the Internet in packets and all users share the same bandwidth. Of course, if every Internet user requested a web page at the same instant, a huge number of packets would be generated and severe congestion would occur. However, by dimensioning the network to reflect the expected number of packets per second, each user can be given bandwidth when they need it without disrupting the service offered to other users. The use of packets to share bandwidth in this way is called "statistical multiplexing".

\*   \*   \*

Router-based data networks such as the one we have just described are likely to cover at least a metropolitan area and some—such as the Internet—can span the globe. However, there is another type of data network that is generally confined to a single building and sometimes to just one or two rooms. This is the Local Area Network, or LAN. LANs are typically used to link personal computers in an office or home environment. LANs may be small, but what they lack in size they generally make up for in performance.

At its simplest, a LAN might allow a group of computers to share a printer or a broadband connection to the Internet. However, many LANs go far beyond this. For example, the LAN can support an "intranet"—a collection of websites that can only be accessed from within an organization. The LAN can also be used by network administrators to ensure that the software installed on each PC is at the correct revision level and is properly licensed. Computer faults can be diagnosed over the LAN and the LAN can be used to distribute software patches and configuration updates for anti-virus software. LANs are such an essential part of modern computing that most new PCs are supplied with one or more built-in LAN interfaces.

LANs first started to appear in the very early days of personal computing. In 1973, Xerox PARC[1] was developing the Alto—an experimental machine that pioneered many of the concepts of modern personal computers. It was recognized that there would be a need to connect these personal computers together and the job was given to a new member of staff called Bob Metcalfe. Metcalfe's solution was to connect each of the computers to a length of coaxial cable,[2] as illustrated in Figure 38.

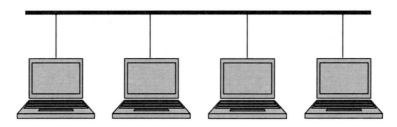

**Figure 38.** Computers linked by coaxial cable.

Any of the computers could transmit a message on the cable and the message would be received by all the computers on the network. The message header would specify the address of the intended recipient, so the other computers would simply discard the message.[3] So far, so good. But what happens if two computers try to transmit at the same time?

One method of solving this problem would be to put a single computer in charge of the network. This master computer would behave rather like the chairman at a meeting, ensuring that each participant is given a chance to have their say and preventing any one speaker from monopolizing the floor. This approach might lead to a very orderly network, but how would the master computer be chosen and what would happen to the network if the master was turned off? Metcalfe chose a more democratic approach that avoided the need for a master computer. Instead of behaving like participants at a meeting, the computers on Metcalfe's network behaved more like guests at a dinner party. Any guest who has something to say simply waits until the previous speaker has finished. If two people start speaking simultaneously, then they both pause and one of them will give way to the other. There is no need for a chairman.

On Metcalfe's network, each computer had a transmitter and a receiver connected to the coaxial cable. If a computer wished to transmit a message, it would start by listening to see whether another computer was using the network. When the network was quiet, it would start transmitting its message but it would use its receiver to check the message as it was transmitted. If the received message corresponded to the transmitted message, then the transmitting computer would know that all was well. If, however, the received message was a garbled version of the transmitted message, then the computer would know that a "collision" had occurred (i.e. another computer had started transmitting at the same moment). In this situation, each computer would stop transmitting and would wait for a period of time before trying again. If the computers waited for the same length of time, then another collision would occur at the end of the pause. For this reason, each computer backed off for a random period of time, making it likely that the first computer to resume transmission would be given uncontested access to the network.

A similar concept—but without collision detection—had previously been used on the Aloha radio network at the University of Hawaii.[4] Since Metcalfe's network was used to link Xerox's Alto computers together, it became known as the Alto Aloha System. This was hardly a catchy title and Metcalfe soon renamed it "Ethernet" (after the luminiferous ether that nineteenth-century physicists believed was necessary to allow light to pass through the vacuum of space).

Xerox began selling Ethernet as a commercial product in 1980, but the development of Local Area Networks was a gradual process until about 1983, when the IBM PC and its various clones started to become widely available. Since then, Ethernet has managed to avoid obsolescence by continually evolving to meet the needs of the market. The original Ethernet design had considerable strengths, but it also had a number of weaknesses. For example, the network would become increasingly inefficient as the level of traffic built up, because more and more time would be wasted recovering from collisions. One solution to this problem was to divide one large LAN into a number of smaller sub-LANs, or "segments". The LAN segments would be linked together by a switch,[5] as illustrated in Figure 39.

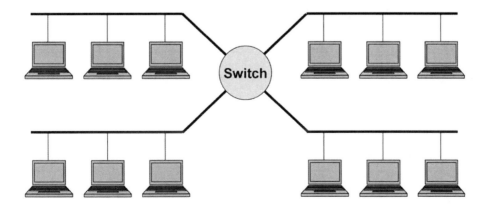

**Figure 39.** LAN segments linked by a switch.

In this arrangement, messages transmitted on one LAN segment are only seen by the switch and by other computers on the same segment—they are not broadcast to all the computers on the network. Messages can be transmitted between computers on the same segment without generating any traffic on other segments. Collisions are also confined to individual segments. If the level of collisions on a particular segment starts to rise, then the problem can be resolved by moving some computers onto a new segment.

The role of the switch in this arrangement is to allow messages to be sent between segments. If a computer on one segment transmits a message to a computer on another segment, then the message will be received by the switch and re-transmitted on the appropriate segment. In order to do this, the switch has to keep track of which computers are located on each segment.[6] The presence of the switch and the segments is effectively invisible to the computers on the network—to them, it appears as if they are all connected to one very big LAN.

The introduction of switching helped to overcome some of the scalability problems inherent in the original Ethernet design and meant that network problems could normally be restricted to a single segment. However, managing the allocation of computers to segments could still be problematic (e.g. when a new computer was required in an area of the building served by a segment that had no spare capacity). The introduction of the switch meant that the network topology was starting to resemble a star, so it was only natural that engineers would consider the possible advantages of a LAN with a pure star topology. This approach is illustrated on the right-hand side of Figure 40.

Each computer now has its own dedicated link to the switch. Since the level of traffic generated by a single computer is far lower than the traffic generated by the computers on a segment, it is no longer necessary to use expensive coaxial cable to carry this traffic—ordinary twisted pair cable can be used instead. Furthermore, the loss of performance caused by collisions can be completely eliminated by using separate pairs of wire for each direction of transmission.

The left-hand side of Figure 40 shows the telephone network found in a typical office environment (the voice switch might be a PABX). The resemblance

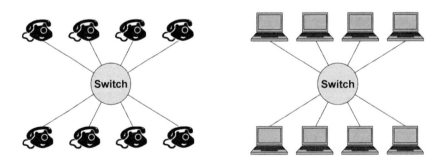

**Figure 40.**  Voice and data network topologies.

between the two networks is not accidental. Both networks use twisted pair cabling and both use a star topology with a switching device at the center. This commonality between the two networks provides opportunities to reduce the cost of wiring within a building. In fact, the need for physically separate voice and data networks within an office environment is starting to disappear as telephones with Ethernet interfaces are added to the LAN.

Since 1980, Ethernet technology has evolved from a "shared media" LAN based on coaxial cable to a "switched" LAN based on twisted pair cable. This has enabled much larger LANs to be built and has improved network performance when carrying high levels of traffic.[7] It has also resulted in lower network installation costs, because cabling can represent a significant proportion of total network costs. Giving each computer its own dedicated link to the switch allows computers to be added or removed from the network without disrupting service to other network users, thereby simplifying network operations. There are also security advantages, because there are no longer any opportunities for computers to read messages that are not addressed to them.

During its long life, Ethernet has successfully seen off competition from a number of alternative technologies, including IBM's token ring LAN. Ethernet is now the undisputed leader in wired LANs, but it is facing increasing competition from "WiFi" wireless LANs.[8] Wireless LANs have the considerable advantage that they avoid the need to install cabling to each computer and so can be implemented very quickly and cheaply. Many laptop computers are supplied with a built-in WiFi interface. However, WiFi networks also have a few weaknesses. To begin with, they operate in unlicensed radio bands, so there is always the risk of interference from wireless networks in an adjacent building.[9] Wireless LANs are also vulnerable to eavesdroppers, so some form of encryption is essential. There have been cases in which people have left chalk marks in the street to indicate the presence of an unprotected wireless LAN in a nearby building—a practice known as "warchalking".

\* \* \*

As we have seen, most Local Area Networks are built around a device called a switch, while most Wide Area Networks are built using devices called routers. At first sight, switches and routers appear to be doing much the same job—they receive a message on one port, check the destination address and then send it on its way via another port. However, switches and routers are fundamentally different. In order to understand this, we need to introduce the concept of network protocols.

A protocol is a language used for communication between different parts of a network. At its simplest, a protocol is nothing more than a set of rules that defines how each part of the network should behave in different situations. Anyone who has listened to the communication between a group of airline pilots and an air traffic controller will have noticed that each participant uses the radio channel according to an agreed set of rules: this is a form of protocol. In a data network, protocols are normally implemented by sending packets between network nodes, but the concept is exactly the same.

Protocols are used in networks for a wide variety of purposes. For example, a protocol can provide flow control to ensure that a transmitter does not send data faster than the receiver can accept it. Another protocol might provide error control to automatically replace data packets that have become lost or corrupted during transmission. Some protocols operate across a single link between two network nodes, while others operate end to end across a network or across a group of connected networks.

Modern network protocols are pretty complex. For this reason, network engineers have adopted a strategy of "divide and rule"—rather than tackling one huge problem, they have divided it up into a number of smaller problems and then solved each smaller problem separately. Experience has shown that a hierarchy of network protocols can be much simpler to implement and test than a single all-purpose protocol. This has led to the development of the OSI[10] seven-layer model, which is illustrated in Figure 41.

Each layer in this "protocol stack" takes services from the layer below, adds its own contribution and then presents an enhanced set of services to the layer above.

Starting at the bottom, we have the Physical layer, otherwise known as "Layer 1". The Physical layer is responsible for managing the physical transmission medium, so it needs to understand the particular characteristics of copper cable, radio or optical fiber. Issues such as synchronization, line codes and modulation schemes are handled by this layer.

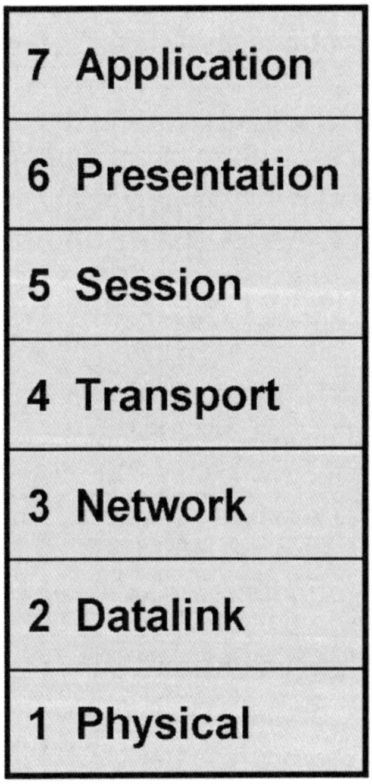

| |
|---|
| **7 Application** |
| **6 Presentation** |
| **5 Session** |
| **4 Transport** |
| **3 Network** |
| **2 Datalink** |
| **1 Physical** |

**Figure 41.** ISO seven-layer model.

Layer 2 (the Datalink layer) does not need to interface directly with the physical transmission medium. Instead, it sends blocks of data down to the Physical layer for transmission to the far end of the line and receives back blocks of data that have traveled in the opposite direction. One of the primary roles of the Datalink layer is to check for errors in each received block of data.[11] In theory, the Datalink protocol could request re-transmission of any data blocks found to contain errors and some early networks worked in this way. However, it turns out to be more effective for the Datalink layer to simply discard corrupted blocks of data and to allow re-transmission to be handled by a higher network layer. This is the approach used on the Internet by the TCP/IP protocol suite.

The Physical and the Datalink layers manage point-to-point links between network nodes.

Layer 3 (the Network layer) is the lowest layer that has a view across the network as a whole. The Network layer has a number of important roles to perform, including the routing of packets. The source and destination of a packet could be at opposite ends of a network—or even on different networks—and it is the responsibility of the Network layer to ensure that the packet is directed towards its intended destination. The routers in modern data networks operate at Layer 3 and below, while Layers 4–7 are implemented in the host computers that are connected to the network.

Layer 4 (the Transport layer) provides reliable transport of data from source to destination. If the Datalink layer has discarded a corrupted block of data, then it is the Transport layer that will request re-transmission of the missing data. The Transport layer also provides flow control to ensure that the transmitter does not transmit data faster than the receiver or the intervening network nodes can accept it. If packets arrive at the receiver in the wrong order, then it is the Transport layer that sorts out the mess.

The remaining layers of the ISO model are really more about applications than about networks. The Session layer establishes and terminates sessions between applications running on different computers, while the Presentation layer provides data conversion facilities such as compression and encryption. At the top of the stack, the Application layer protocols enable the user to carry out tasks; for example, the File Transfer Protocol (FTP) can be used to transfer data from one computer to another.

When two computers are communicating across a network, each layer in the protocol stack is effectively running its own separate conversation, as illustrated in Figure 42. Let's suppose that files are being transferred from one computer to another, so an Application layer conversation will be taking place between the two computers. Immediately below this, the Presentation layers may be encrypting the files prior to transmission and then decrypting them again when

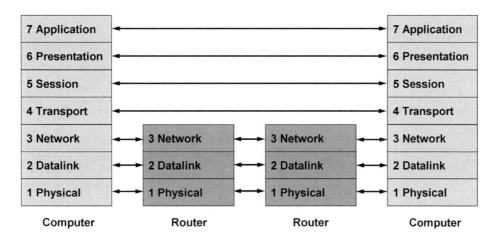

**Figure 42.** Computers communicating across a network.

they arrive. Below this, the Session and Transport layers will each be having their own end-to-end conversations. For the bottom three layers of the stack, the conversations take place on a link-by-link basis rather than end-to-end. Each computer talks to an adjacent router and the routers then talk to other routers. The diagram above shows communication between two computers that are linked by two routers. However, for computer–computer connections that span multiple networks, the number of intervening routers could obviously be much larger than this.

People are often confused by diagrams of this type because they appear to show a direct physical link between the top four layers of the protocol stack. How can this be, when the intervening routers can only handle Layers 1–3? The answer, of course, is that there is some multiplexing going on here. The simultaneous conversations that are going on at each layer in the protocol stack are all sharing the same cables or radio links in the Physical layer. An example may help to clarify the situation.

Let's suppose that a user needs to download a file from a computer located somewhere on the network. To start with, the Layer 7 application on the destination computer needs to talk to its opposite number on the source computer. However, the application does not need to know how its messages are transmitted across the network—it simply needs to be sure that the messages will be delivered quickly and reliably. It therefore passes each message down to the Presentation layer, secure in the knowledge that the layers below it in the protocol stack know how to deliver the message. The Presentation layer takes the message and performs any necessary operations on it before passing it down to the Session layer (remember that each layer can only pass messages to the layer immediately above or below it). In this way, the message ripples down through the stack until it reaches the Physical layer.

As it passes down the layers of the stack, the message acquires additional protocol bits. Each layer adds a wrapper around the data that it received from the

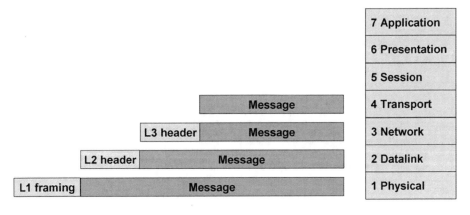

**Figure 43.** Message in nested protocol wrappers.

layer above, so the protocols end up nested inside each other like a Russian doll. For example, each message that is passed down from Layer 4 to Layer 3 is inserted into the payload of a Layer 3 packet. Similarly, when the Layer 3 packet is passed down to Layer 2, another header is added. This process is illustrated in Figure 43.

When the message reaches the Physical layer, it sets off on its journey across the network. Let's consider what happens when it encounters its first router. The Physical layer in the router will strip off the Layer 1 framing and synchronization bits to recover a block of data, which is then passed up to the Datalink layer. Here, the Layer 2 header is removed to reveal the Layer 3 packet, which is then passed up to the Network layer. The Layer 3 packet header carries the routing information that allows the router to decide where the packet should go next. When this routing decision has been made, new Layer 2 and Layer 1 headers and framing are added and the message is once again sent on its way. This process is repeated at each router that the message encounters until, finally, it arrives at its destination. Here, the message is passed up the stack from one layer to the next and the various layers of overhead are stripped off. Finally, with suitable fanfare, a beautiful error-free message is presented to the Application layer running on the destination computer. If a response is required, then it will be passed down from the Application layer to the Presentation layer and the whole sequence of events will be repeated in the opposite direction.

* * *

Over the years, many different protocol suites have been developed to implement the layers in the OSI model. Some—such as IBM's SNA and Apple's AppleTalk—were developed by particular equipment vendors to operate primarily or exclusively with their own products. Others—such as X.25 and ATM—were developed by international standards bodies and were intended to promote interoperability between equipment from different vendors.[12] Some key Internet protocols—such as TCP and IP—were initially developed as part of research activities but have subsequently been recognized as standards.

Some of the differences between protocols stem from philosophical disagreements about how intelligent a network should be. If functions such as error correction and flow control are provided by the network,[13] then it is reasonable to assume that the network will also provide more advanced features. This leads on to the concept of an "intelligent network". Network operators tend to like the intelligent network idea because advanced features allow them to differentiate their services from competing offerings and thereby justify higher prices. However, the Internet has demonstrated that a "dumb network" can also have considerable advantages. The physical infrastructure of the Internet provides very limited functionality—its routers do not guarantee that a sequence of packets will arrive in the correct order or even that they will arrive at all. Error correction and flow control are provided by the TCP protocol but TCP is implemented by computers connected around the edges of the network rather than by the network itself. Even if a massive range of advanced services and features are created in this way, the network itself remains as dumb as it always was.[14]

A major advantage of the dumb network approach is that anyone—not just the network operator—can build new features to enhance the capabilities of the network. Important applications such as the World Wide Web and the Skype telephone service have been built on top of the basic transport services provided by the Internet. These applications were not developed by network operators and they did not require any changes to be made to the network infrastructure. Applications evolve over time, but the dumb network is very unlikely to become obsolete because the basic functionality required to transport bits from one place to another is unlikely to change. If new applications are required, then enterprising individuals or organizations will develop them. As the Internet has shown, dumb networks can promote rapid innovation.

\*   \*   \*

Earlier in this chapter, we raised the question of why switches are different from routers. We are now in a position to answer this question. In a nutshell, switches operate up to Layer 2 in the ISO model, while routers operate up to Layer 3. To illustrate this point, let us suppose that a computer on a LAN wishes to communicate with a computer located somewhere else on the Internet. The situation is illustrated by the dotted line in Figure 44.

The computer on the LAN will generate a Layer 3 packet with higher-layer protocols hidden inside it. This IP packet will be encapsulated in an Ethernet frame (the Layer 2 equivalent of a packet) and sent to the LAN switch. The switch is a Layer 2 device; it understands how to handle the Ethernet frame, but it has no understanding of the IP packet that is hidden inside the frame (or, indeed, of the higher-level protocols that are hidden inside the IP packet). The switch will pass the Ethernet frame to the router on the LAN that has a connection to the Internet. This router strips off the Ethernet frame to reveal the IP packet inside. Now, the router is a Layer 3 device, so it knows how to handle an IP packet. The packet will be passed across the Internet from one router to the next (in the manner described at the start of this chapter) until it reaches its destination.

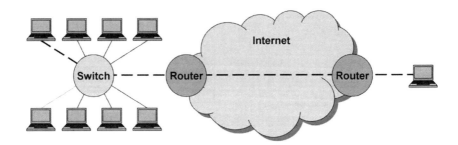

**Figure 44.**   Switches and routers.

Although we have now introduced a fundamental difference between switches and routers (switches operate at Layer 2 in the protocol stack; routers operate at Layer 3), we have not yet explained why both types of device are required. Why, for example, could we not build the Internet using switches instead of routers? The answer lies in the differences between Layer 2 and Layer 3 addresses. If you go to your local computer store and buy a laptop computer with a built-in LAN interface, you will probably find a sticker on it somewhere that says something like "MAC Addr: 00:C0:49:D7:16:81". This is the Layer 2 address of the device. It is "burnt" into the device hardware during manufacture and it never changes. The Layer 2 address format can support more than 2,800 billion billion addresses, so we can safely allocate a unique address to every LAN device in the world without any risk of running out of addresses!

When you connect the device to a switched LAN, the switch will make a note of the device's Layer 2 address in a look-up table, along with details of the switch port to which it is connected. Every time a message arrives at the switch, the Layer 2 destination address in the message header is compared with the addresses stored in the switch's look-up table. If a match is found, then the table entry will indicate the switch port to which the message should be sent. In this way, the switch can build up a record of how to find each computer on its network.

It is worth thinking a bit more closely about this look-up table. One way to implement it would be to build a look-up table with one entry for every possible Layer 2 address, but a table with over 2,800 billion billion entries is going to occupy a gigantic amount of memory! Since the number of devices on a large LAN might be measured in hundreds, a much more practical solution is to construct a table with (say) 1,000 entries and to use one table entry to store the address of each device connected to the LAN. Such a table could be stored in a very modest amount of memory. If the table entries were then sorted into numerical order, it would be a relatively simple task to check whether a particular address was held in the table.

Now let's suppose that the Internet was built using switches rather than routers. The number of users connecting to the Internet on a regular basis is measured in billions, so each switch would be expected to maintain a massive look-up table. Even if it was practical for each switch to keep track of every device on the Internet, the cost of the memory and processor power required to store and manage this huge table would be prohibitive. So, how do routers get around this problem? The answer is that they operate using Layer 3—rather than Layer 2—addresses. Layer 3 addresses are allocated hierarchically, so devices on a particular LAN would usually be allocated addresses from the same Layer 3 address range. This means that a router does not normally need to read the whole of a Layer 3 destination address when routing a packet—it simply needs to consider enough digits to determine the general area to which the packet is heading and then send it off in the right direction. As the packet approaches its destination, the routers will have to start reading a larger proportion of the address and the final router will need to read the whole of the address.

It's much the same in the postal system. When a letter arrives at the first sorting office, it can be sorted by simply reading the destination county. When it arrives in the right county, it can then be sorted by town. When it arrives in the right town, it can be sorted by street. At each stage, another line in the address needs to be considered. An Internet built of switches rather than routers would be rather like a postal system that only used house names to address letters (assuming that no two houses could have the same name). Letters could be addressed simply to "Cedar Cottage" or "Mon Repos"; there would be no possible ambiguity over their destinations, but the postal system would grind to a halt. By having a hierarchical address structure, we avoid the need for every postman to know the location of every home and business in the country (or, indeed, the world).

From a logical point of view, an Internet built entirely of switches rather than routers would be one massive LAN. Even if we could manage the problems caused by non-hierarchical addresses, we would run into major difficulties with broadcast messages. As the name implies, broadcast messages must be delivered to every device on the LAN and they are used by certain network protocols in situations in which the required destination address is not known. For example, if a new computer is added to a LAN, it might use a broadcast message to announce its arrival to the other computers on the LAN. It is not hard to imagine the problems that would be caused if every broadcast message had to be delivered to every device on the Internet. However, if a large LAN is divided into two smaller LANs linked by a router, then broadcast messages will only be delivered to the devices on one LAN or the other—they will not make it through the router. If we further sub-divide each of the LANs into a number of smaller LANs linked by routers, then the broadcast messages will be even more constrained. Eventually, this process leads to a situation in which a very large number of small LANs are linked together by a network of routers. In simple terms, that's pretty much what the Internet looks like today!

# 9    The birth of the Internet

After a major advance in technology, it usually becomes apparent that there were a number of people working independently on the same concept at about the same time. This was certainly the case with the development of packet switching. With the benefit of hindsight, it is now clear that there were at least three separate groups working in parallel—but completely independently—on the same basic idea. These groups were located at the RAND Corporation in California, at the National Physical Laboratory (NPL) in London and at the Massachusetts Institute of Technology (MIT). As a result of the overlapping achievements of each of these groups, the question of "Who got there first?" is even more heavily disputed in the case of packet switching than it was in the case of the telephone or the telegraph.

Some of the drivers behind packet switching came from US military requirements. In the early 1960s, the Cold War was at its height and the Cuban missile crisis had demonstrated the appalling dangers of the nuclear stand-off between the United States and the Soviet Union. The doctrine of Mutual Assured Destruction, which developed at about that time, required an attacker to believe that any advantages that might be gained by initiating a nuclear exchange would not be sufficient to justify the retaliation that would inevitably follow. On the other hand, the temptation to initiate a nuclear attack (a "first strike", in the jargon) would be greater if the enemy could be left so devastated that they would be unable to retaliate.

For this reason, both sides sought to develop a "second strike" capability that would enable them to respond in spite of the damage caused by an initial attack. This gave rise to some real concerns in the United States. It was pointed out that short-range Soviet missiles could take out all the United States' strategic air command bases near the Soviet Union within a period of just a few minutes. Furthermore, the centralized switching nodes of the US telephone network made it highly vulnerable to attack and one nuclear explosion at high altitude could prevent HF radio communication for many hours.[1] To retain a viable second-strike capability, the command, control and communication systems would have to remain operational after the first strike.

This problem came to the attention of a young engineer called Paul Baran, who worked in California for the RAND Corporation.[2] In a series of papers published between 1962 and 1964,[3] he developed his concept of a "network of unmanned digital switches implementing a self-learning policy at each node, without the need for a central—and potentially vulnerable—control point".[4] Rather than spending huge sums of money trying to protect critical parts of the

A. Wheen, *From Dot-Dash to Dot.com: How Modern Telecommunications Evolved from the Telegraph to the Internet*, Springer Praxis Books, DOI 10.1007/978-1-4419-6760-2_10, © Springer Science+Business Media, LLC 2011

network infrastructure, Baran proposed a much cheaper solution: a network that could sustain considerable damage and continue to operate effectively.

Baran's thinking was incredibly far-sighted. Not only did he describe all the basic features of a packet switching network, but he also addressed the Quality of Service issues associated with carrying voice on the network.[5] He even included "provisions to allow each user to 'carry his telephone number' with him"—a facility known today as "roaming". Furthermore, the resilience inherent in Baran's design meant that it could be constructed using cheap and relatively unreliable components. Amazingly, his cost estimates suggested that the network would be *two orders of magnitude* cheaper than analog networks that were being proposed by the military.

Not surprisingly, some of the fiercest opposition to Baran's proposals came from AT&T, which, at that time, held an effective monopoly on long-distance telephone services in the United States. To AT&T executives, the idea of building an all-digital network probably seemed dangerously radical but the fact that it discarded the circuit switching concept that had been at the heart of telephone networks since the days of Alexander Graham Bell made it totally unthinkable. Baran remembers one outburst from a senior AT&T executive after a particularly difficult meeting: "First, it can't possibly work, and if it did, damned if we are going to allow the creation of a competitor to ourselves."[6]

In 1965, RAND Corporation formally recommended that the US Air Force should construct an experimental version of Baran's network. Although this proposal was enthusiastically accepted by the Air Force, it ran into problems in Washington. Robert McNamara, the Secretary of State for Defense, had decided to consolidate the long-distance communication requirements of the three armed services, so the Air Force was no longer able to make its own decisions on this matter. Furthermore, the Defense Communications Agency that had been set up to manage these issues was staffed mainly by ex-AT&T people with little or no understanding of packet switching. If the project went ahead, it would be these people who would be responsible for building the network and they would almost certainly fail. If this happened, then the whole concept of packet switching would be seriously damaged and funding for future projects would be very hard to find. Baran faced an agonizing choice: pursue the project and run the risk of killing the concept, or abort the project and hope that a better opportunity would come along in the future. Reluctantly, he decided to abort the project.

Baran's work may have reached an impasse, but packet switching was not dead. Although Baran was unaware of it, there was a group at the National Physical Laboratory (NPL) in London who were working on similar ideas. Donald Davies had started work on an improved method of enabling computers to communicate over telephone lines in 1965 and he outlined his ideas on "The Future Digital Communication Network" in a public lecture in March 1966. At the end of the lecture, he was approached by an official from the Ministry of Defence, who told him about Paul Baran's work at RAND. It subsequently turned out that there were some striking similarities between the two concepts,[7] even though they had been developed entirely independently. Whilst there is no dispute that Baran's work preceded Davies's, it was Davies's term "packet switching"—rather than Baran's "distributed adaptive message block switching"—that was ultimately adopted for the new technology.

\* \* \*

J. C. R. ("Lick") Licklider was a visionary. In 1962, he wrote a series of articles describing a "Galactic Network"[8]—a global network of computers that would allow programs and data to be accessible from any site. Although the concept was tongue-in-cheek, his ideas proved to be prophetic. Many of the features that he described can be found in today's Internet.[9]

In 1963, Licklider took up a position with the US Department of Defense's Advanced Research Programs Agency (ARPA).[10] ARPA had been created in 1958 as part of the US response to the launch of Sputnik and its mission was to create leading-edge technology for the US military. Although Licklider left ARPA before his ideas had been put into practice, he set in train a sequence of events that led to the development of the world's first wide-area packet switching network: the "ARPANET".

Despite their close military connections, ARPA were not trying to build a command and control system that would survive a nuclear attack (indeed, during the early stages of the ARPANET design, the engineers involved were not even aware of Paul Baran's pioneering work). The primary reason for building the network was to enable ARPA-sponsored researchers to make use of ARPA-funded mainframe computers in different locations.[11] As a former director of ARPA put it, "it came out of our frustration that there were only a limited number of large, powerful research computers in the country, and that many research investigators who should have access to them were geographically separated from them".[12]

Responsibility for creating the ARPANET was assigned to Lawrence Roberts, a young computer scientist who joined ARPA in 1966. Roberts had been interested in computers since his undergraduate days at MIT. During his subsequent studies for a Master's degree (on data compression) and a PhD (on the perception of three-dimensional solids), he seems to have found time to become an extremely proficient programmer. He first became interested in computer networks in 1964 when he attended a conference at Homestead, Virginia, at which Licklider was also present. As he recalled during an interview in 1989:[13]

> "We had all of these people doing different things everywhere, and they were all not sharing their research very well ... we had to do something about communications ... the idea of the galactic network that Lick talked about, probably more than anybody, was something that we had to start seriously thinking about."

The original ARPANET concept was to link the various computers together via the telephone network and to use the computers themselves to switch the messages around. However, at a meeting in early 1967, many potential participants complained that the demands of the network would use up a significant amount of their available computer power. One of the participants, Wesley Clark, suggested that the communication links to each site should be managed by a separate computer, so that the impact on existing computers would be minimal. This turned out to be an inspired idea. Small but powerful minicomputers were just starting to appear. The network routing software could be written to run on just one type of minicomputer and did not have to take account of the differences between the mainframe computers that were being linked by the network. This would save on programming costs and facilitate network upgrades. It also gave ARPA complete control over the operation of the network. Clark's minicomputers became known as Interface Message Processors—or IMPs for short. They were the forerunners of today's routers.

There is potential for confusion here: we have to distinguish between computers that are used to handle networking functions (i.e. the IMPs) and computers that are providing services to end users. In order to be clear, we will now adopt Internet terminology and refer to the latter as "host computers", or simply as "hosts".

Larry Roberts presented a "Plan for the ARPANET"—including the IMP concept—to a seminar in Gatlinburg, Tennessee, in October 1967.[14] However, the striking lack of detail in the presentation suggests that Roberts had very little idea how the network would actually work. Fortunately, the seminar was also attended by Roger Scantlebury from Donald Davies's team at NPL. Scantlebury presented a paper setting out the packet switching concept in detail. His paper also described the design of a network that NTL were proposing to build to link together 10 host computers and a large number of peripheral devices. Amazingly, it appears that this was the first time that the ARPA project team had encountered the concept of packet switching and they recognized its importance immediately. Scantlebury had quoted Baran as a reference and the ARPA team soon tracked down Baran's RAND papers when they returned to Washington. The threads were starting to come together.

Some years later, Baran compared the process of technological innovation to the building of a cathedral:

> "Over the course of several hundred years, new people come along and each lays down a block on top of the old foundations, each saying 'I built the cathedral'. Next month, another block is placed atop the previous one. Then comes along an historian who asks, 'Well, who built the cathedral?' Peter added some stones here, and Paul added a few more. If you are not careful, you can con yourself into believing that you did the most important part. But the reality is that each contribution has to follow onto previous work. Everything is tied to everything else."[15]

In August 1968, Roberts released a specification for the ARPANET in which he set out how the IMPs would work. Each host computer on the network would be connected to an IMP via a serial interface. The IMP would break down the data from the host into packets. Source and destination addresses would be added to each packet, along with additional bits to allow the receiver to check for errors in transmission. The IMP would then choose the best route for each packet and would send the packets on their way down leased telephone lines running at 50 Kbit/second. The IMPs at intermediate sites would provide store-and-forward functions to ensure that the packets reached their intended destination. When all the packets had arrived, the headers would be stripped off and the data would be assembled to recreate the original message.

Roberts's IMP specification was sent to 140 potential suppliers. Most of them regarded the idea as ridiculous and only 12 proposals were actually received. The contract was eventually awarded to Bolt, Beranek and Newman (BBN), a small computer company based in Cambridge, Massachusetts.[16] Despite their modest size, BBN employed a world-class team of scientists and engineers drawn from the nearby universities of Harvard and MIT. The team included Robert Kahn, who, as we will see, was destined to play a starring role in the subsequent development of the Internet.

It didn't take BBN long to achieve results. The first ARPANET node was installed in September 1969[17]—only 9 months after the contract was awarded. BBN based their initial design for the IMP on a ruggedized version of the Honeywell DDP-516 minicomputer. Each IMP could be directly connected to up to four host computers and could support network links with up to six remote IMPs.[18] The first four network nodes were operational by December 1969 and the network expanded to 13 nodes during the course of the following year.

This was a major achievement, but it was not enough to enable successful networking. If a host computer transmitted a packet into the network, then the IMPs would duly deliver it to the destination host. But without appropriate software, the destination host would have no idea what to do with the packet. Indeed, without appropriate software, the source host would never have generated the packet in the first place! In order for the host computers to make use of the network, they needed to learn how to talk to each other.

This may seem obvious today, but things were very different in 1969. Computers in those days were physically large and very egocentric. They were used to issuing orders to peripheral devices such as printers, but they had no experience of talking to each another as equals. Before that could happen, a common language for host computers would have to be found and it was far from clear how this common language should be designed. Since the network was required to link many different types of host running incompatible software, the ARPANET designers faced a formidable challenge.

\* \* \*

In the previous chapter, we introduced the ISO seven-layer model as the standard structure for a protocol stack. Of course, the ISO model is based on many years of experience and very little of this experience existed when the

ARPANET was built. The bottom three layers of the ARPANET protocol stack were to be implemented in the IMPs, so the responsibility for developing these layers fell to BBN.[19] The remaining four layers of the stack would run only in the host computers connected to the network and responsibility for developing these protocols was given to the Network Working Group under the leadership of Steve Crocker. Their task was to create a host protocol that was powerful enough to support all the types of communication that were likely to occur, while remaining sufficiently simple to allow implementation on all the different types of host computer that would be attached to the network. As Steve Crocker later recalled:[20]

> "Over the spring and summer of 1969 we grappled with the detailed problems of protocol design. Although we had a vision of the vast potential for intercomputer communication, designing usable protocols was another matter. A custom hardware interface and custom intrusion into the operating system was going to be required for anything we designed, and we anticipated serious difficulty at each of the sites. We looked for existing abstractions to use. It would have been convenient if we could have made the network simply look like a tape drive to each host, but we knew that wouldn't do."[21]

The operating system on each host computer would have to be modified to support basic network operations such as establishing a connection with a remote host, sending and receiving data over that connection and then terminating the connection. This was not going to be simple. As Crocker put it:

> "Unfortunately, operating systems of that era tended to view themselves as the center of the universe; symmetric cooperation [with other hosts] did not fit into the concepts currently available within these operating systems."[22]

The Network Working Group were also required to develop protocols that would use the Network Control Program to perform specific tasks. It was agreed that the first two applications to be addressed would be remote log-in (the ability to run an interactive session on one host while physically connected to another) and host-to-host file transfer. These protocols later became known as Telnet (TELetype NETwork) and FTP (File Transfer Protocol), respectively. The initial version of the Network Control Program[23] was completed in December 1970 and roll-out across the ARPANET occurred during the following 2 years.[24]

The development of the Network Control Program was a major step forward, but the extraordinary growth of the ARPANET inevitably exposed some of its limitations. For example, it was becoming clear that the ARPANET could not continue to exist in splendid isolation; just as the telephone network in one country must be able to handle telephone calls to networks in other countries, so a data network must be able to exchange packets with other data networks. Unfortunately, however, the Network Control Program could only address host computers that were directly connected to the ARPANET, so machines on other networks were effectively out of reach.

Even without this problem, differences in design would have made it difficult to link the ARPANET with other contemporary data networks. These differences were not simply caused by a lack of agreed standards, although that was certainly a factor—they also reflected the different forms of transmission used to link network nodes together. For example, the SATNET network used satellite links and so the protocols had to accommodate much longer transmission times that would occur on the cable-based ARPANET. Similarly, the protocols used on PRNET (Packet Radio Network) had to cope with the high packet loss rates that could occur on mobile, radio-based nodes.

It was a former member of BBN's IMP team who set himself the task of finding a logical architecture that would enable networks with different characteristics to be linked together into an "Internet". Before joining BBN, Robert Kahn had been a professor of electrical engineering at MIT, where he had worked on mathematical issues in communication and information theory. In 1972, he had organized the first public demonstration of the ARPANET's capabilities at the International Conference on Computer Communications in Washington, DC. This had included "conversations" between a computer at MIT that was programmed to behave like a psychiatrist and a computer at Stanford that mimicked the behavior of a paranoid psychotic. One of their conversations went as follows:

Doctor: "Good evening. Tell me your problems."
Patient: "People get on my nerves sometimes."
Doctor: "I am not sure I understand you."
Patient: "You should pay more attention."
Doctor: "Suppose you should pay more attention."
Patient: "You're entitled to your own opinion."
Doctor: "What makes you think I am entitled to my own opinion?"
Patient: "It bothers me just to be around people in general."

These demonstrations of long-distance networking may not have done much to advance the treatment of paranoid psychosis, but they brought the ARPANET's capabilities to a much wider audience. They also helped to silence the doubters who had claimed that packet switching would never work.

In the following year, Robert Kahn moved from BBN to ARPA to work for Larry Roberts. He soon identified four key characteristics of the new Internet architecture that he was seeking:[25]

- Networks should not be required to make internal design changes before they could connect to the Internet. This was a pragmatic requirement that reflected the difficulty of forcing all data networks to adopt a common design.
- Packet transport across a network would be on a best-effort basis and would not be guaranteed. If a packet did not make it to the final destination, then

it would be retransmitted by the source host. This was a significant departure from the design of the ARPANET, where the network nodes (i.e. the IMPs) rather than the hosts (the Network Control Program) were responsible for providing end-to-end reliability.

- Gateway devices would be used to connect the networks together and to overcome any incompatibilities between them. A gateway would behave like a host so far as each of its connected networks was concerned. In order to keep things simple, a gateway would not be required to retain any status information about the packets flowing through it.
- There would be no global control—the Internet would operate by collaboration rather than command. This approach helped to reinforce the free-wheeling and slightly anarchic culture that became a key factor in the subsequent development of the Internet.

It had been necessary to modify the Network Control Program to run on each different type of host computer that was connected to the ARPANET and Kahn realized that the same requirement would apply to any new host protocol that he defined. This would be a major undertaking. He needed to make sure that his new protocol could be implemented efficiently on all the different operating systems that hosts would be running. He therefore invited Vint Cerf, who had been involved in the design and development of the Network Control Program, to help him in his work. Kahn and Cerf had first met in early 1970 while conducting tests on the embryonic ARPANET. At that stage, Cerf had been a postgraduate student at UCLA, while Kahn had been working for BBN. In 1972, the International Network Working Group had been formed to coordinate various American and European networking initiatives and Cerf had been appointed its first Chairman. Later that same year, Cerf had taken up an assistant professorship in computer science and electrical engineering at Stanford University.

After several months of discussion, Kahn and Cerf arrived at the concept illustrated in Figure 45. This diagram is based on one that, according to Internet folklore, was sketched by Cerf on the back of an envelope while waiting in a hotel lobby. It shows the SATNET and PRNET networks linked to the ARPANET by gateways. The SATNET, PRNET and ARPANET had different interfaces, different transmission rates and different maximum packet sizes, but the gateways would sort out these problems. As a result, a host on one network could communicate with a host on either of the other networks, even if—as in the case of PRNET and SATNET in the diagram above—the two networks were not directly linked. This architecture allowed a large number of independent networks to operate together as a single logical network. It is a concept that lies at the very heart of the Internet.

In September 1973, the first documented description of the architecture was presented to a special meeting of the International Network Working Group at a conference at the University of Sussex in Brighton. The architecture was further refined during the course of this meeting and was subsequently published in May

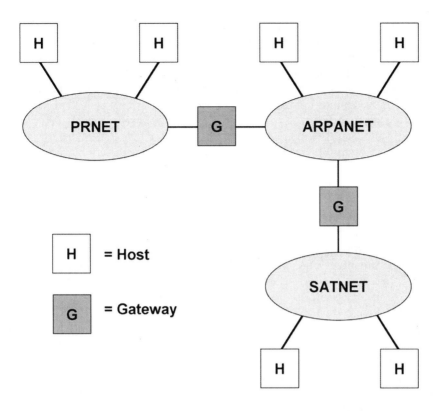

**Figure 45.** Internet concept.

of the following year.[26] This landmark paper described a new protocol called TCP (Transmission Control Protocol). ARPA placed a contract with BBN to develop a prototype version of the gateway and Cerf's group at Stanford were contracted to produce the initial design of the TCP software. ARPA subsequently contracted with BBN and with University College, London, to develop independent implementations of the TCP software to run on different host computers. The three teams collaborated during the testing of the software, with Cerf providing overall leadership. The first demonstration of TCP operating across ARPANET, SATNET and PRNET through BBN-supplied gateways occurred in July 1977.

Kahn recognized that different forms of traffic might require different variants of TCP to transport them properly. Applications such as file transfer required error-free transmission of the data, so occasional delays caused by the need to re-transmit lost or corrupted packets would be acceptable. If, on the other hand, the network was carrying a telephone conversation, the requirement would be completely different—you cannot ask someone to stop talking while a packet is re-transmitted! In this case, it is better to accept that packets will occasionally be lost or discarded, resulting in a (probably imperceptible) break in transmission.

The initial implementations of TCP focused on the error-free transmission of data, because this was more immediately useful. However, in March 1978, TCP was split into two protocols: a slimmed-down version of TCP that handled issues such as errors and flow control, and a separate protocol called IP (Internet Protocol) that would handle the lower-level functions associated with forwarding packets across a network. For applications such as voice, the User Datagram Protocol (UDP) would be used instead of TCP. Today, TCP and IP are so frequently used together that they are referred to collectively as TCP/IP.

In 1980, the TCP/IP protocol suite was adopted as a standard by the US Department of Defense. It was subsequently decided that all ARPANET nodes would switch from the old Network Control Program to TCP/IP. This was not a simple task, because the ARPANET hosts would have to convert simultaneously. After several years of planning, the cut-over finally occurred on January 1, 1983. Things went surprisingly well and participants proudly wore badges displaying the message "I survived the TCP/IP transition".

At about this time, the Pentagon was becoming concerned about civilian and military communications sharing the same network facilities. Once the ARPANET had migrated from NCP to TCP/IP, it became possible to separate out the US military components of the ARPANET into a separate network called MILNET. Sixty-eight of the ARPANET nodes moved across to MILNET, while the remaining 45 nodes remained with the ARPANET and continued to support research needs. In order to allow the exchange of email, the MILNET and ARPANET remained connected via a small number of gateways. These gateways could be quickly disconnected if this ever became necessary for security reasons.

In the late 1980s, the National Science Foundation (NSF) decided to create supercomputer centers at five high-profile American universities. The small number of centers reflected the high cost of setting them up, but access to these valuable resources could not be restricted to just five universities. After unsuccessful discussions with the operators of the ARPANET, NSF decided to build their own network (NSFNET) based on ARPANET technology. The five supercomputer centers were linked by 56 Kbit/second lines to form a core network, with other sites being connected together in chains to form regional networks radiating out from the core. NSF funded connections to new sites on condition that these sites would allow other new sites to connect to them, thereby promoting the growth of the regional networks.

By 1990, NSFNET had replaced ARPANET as the core of the Internet. The number of computers connected to NSFNET was now far greater than the number connected to ARPANET and NSFNET used transmission lines that were more than 25 times faster. The ARPANET was 20 years old and it was showing its age. One by one, the remaining host computers were moved across to NSFNET, the IMPs were de-commissioned and the ARPANET passed into history. In some ways, it was rather like the passing of an elderly and much loved parent, but it left behind a vigorous and growing Internet family.

The early years of the Internet had been funded by US government investment and government rules did not allow use of the Internet for commercial activities. However, by 1988, it was becoming clear that this restriction might seriously

Cerf and Kahn receiving the Presidential Medal of Freedom.[28]

hamper the growth of the Internet. Email had been one of the major successes of the ARPANET project and the idea that commercial and scientific users would have to communicate via physically separate networks was becoming increasingly untenable. Permission was therefore given to connect the Internet with the commercial MCI Mail email system as part of an email interconnection experiment. This interconnection was completed in the summer of 1989 under the direction of Vint Cerf.

In 1991, the NSFNET Acceptable Use Policy was modified to allow use by "research arms of for-profit firms when engaged in open scholarly communication and research". Two years later, the US Congress finally passed legislation that allowed NSFNET to be opened up for full commercial use. Commercial ISPs were spun out from the NSFNET regional networks (creating such companies as UUNET and PSINet) and Network Access Points were established in New York, Washington, DC, Chicago and California to facilitate interconnection between ISPs. By 1995, the commercial side of the Internet was firmly established and NSF was able to conclude that its financial support for NSFNET was no longer required. The Internet had grown up and had learned how to pay its way in the world.[27]

\* \* \*

Many of the people associated with the early development of the ARPANET could reasonably be described as visionaries. However, none of them could possibly have foreseen just how significant their work would prove to be. Within a few years, packet switching evolved from a research topic into a proven networking technology. Networks based on Internet technologies are now displacing telephone networks in much the same way as telephone networks displaced the telegraph networks that preceded them.

As we have seen, the initial development priorities for ARPANET applications were file transfer (the FTP protocol) and a facility to enable logging on to a computer from a remote location (Telnet). This choice reflected ARPA's objectives for the network, which were to enable wider use of the large and expensive computers that they were funding, and to promote the sharing of information among ARPA-funded researchers. However, the explosive—and totally unexpected—growth of email on the ARPANET provided the first indications that applications for this technology would stretch far beyond basic computer networking. Indeed, more recent developments have taken Internet applications into realms that could hardly have been imagined even a few years ago and the pace of development is showing no signs of slowing. It is to Internet applications that we must now turn our attention.

# 10 Life in cyberspace

The Internet is a "dumb" network. IP routers do their best to move packets from source to destination, but there is no guarantee that the packets will arrive in order or even that they will arrive at all. It is the computers sitting around the edges of the network that make the network operate reliably and respond intelligently to users. Some of these computers are used to host websites. Others are supporting email, and podcasts, and telephony, and networked computer games—and a whole spectrum of other applications ranging from chat rooms to virtual worlds. In this chapter, we introduce some of the most important applications to appear on the Internet.

## 10.1 Email

The development of email was one of the major successes of the ARPANET program. Email systems had existed since the early 1960s, but they were restricted to sending messages between users on the same mainframe computer. However, in March 1972, Ray Tomlinson at BBN cobbled together an existing email program with an experimental file transfer program. The software that he produced enabled network developers to send and receive messages between different computers on the ARPANET.

Expanding the scope of email from a single computer to a network of computers required a change in the email address format. If email was confined to a single computer, a valid username was sufficient to specify the destination; once email could be sent between computers, it became necessary to specify not only the username, but also the name of the host computer. Spaces could not be allowed in an email address, so Tomlinson had to find a symbol to separate the hostname from the username within the address. He eventually chose the @ character, thereby creating the username@hostname email address format that is still in use today. As he later explained, "there was nobody with an @ sign in their name that I was aware of".[1,2]

Tomlinson's email software was a groundbreaking development but it lacked many of the basic features that we take for granted in modern email systems. For example, there was no "reply" function, so a new message had to be created from scratch in order to send a response. Furthermore, received messages were stored as one continuous stream of text, so finding a new message could be tedious. Within 3 months, Larry Roberts had written an email program that could list, selectively read, file, forward and respond to messages. At this point, email usage

A. Wheen, *From Dot-Dash to Dot.com: How Modern Telecommunications Evolved from the Telegraph to the Internet*, Springer Praxis Books, DOI 10.1007/978-1-4419-6760-2_11,
© Springer Science+Business Media, LLC 2011

on the ARPANET really took off. A study conducted during the following year found that email represented 75% of all network traffic. Today, email is still one of the most important and widely used applications on the Internet.

Email is an extremely efficient means of communication. It is dramatically cheaper and quicker than sending a letter, while still providing adequate documentary evidence to support many business transactions. Email has also displaced telephone calls in situations in which it is difficult for two people to find a convenient time to communicate—perhaps because they both have busy schedules or because they are living in different time zones. Using email, a question can be sent from the United Kingdom at the end of the working day and an answer will be waiting by the following morning—courtesy of colleagues in Los Angeles.

However, email cannot match a telephone conversation for communicating emotion and feeling. In 1979, Kevin MacKenzie suggested that -) could indicate that a sentence was tongue-in-cheek. (To view this properly, turn the page clockwise so that the curved bracket represents a smile. The hyphen represents the tongue in cheek.) However, the idea did not really catch on until 1982, when Scott Fahlman proposed that :-) could be used to indicate a joke, with :-( indicating that something was NOT a joke. This appealed to the playful spirit of Internet users and hundreds of these "emoticons" were created. A few examples are shown in Table 6.

**Table 6.** Emoticons.

| | |
|---|---|
| ;-) | Wink |
| :-* | Kiss |
| >:-0 | Anger |
| :-*) | Blushing |
| 0:-) | Innocence (halo over head) |

Not surprisingly, modern software packages have taken this idea a step further. If you type :-) into Microsoft Word, it will render it as ☺.

## 10.2 World Wide Web

The World Wide Web is such a fundamental part of the Internet that many people mistakenly believe that it *is* the Internet. The web was invented in the early 1990s by a young Englishman called Tim Berners-Lee, who was working at the CERN atomic research centre near Geneva. As he later recalled:

> "The real world of high-energy physics was one of incompatible networks, disk formats, data formats, and character-encoding schemes, which made any attempt to transfer information between computers generally impossible."[3]

In order to gain maximum benefit from the results of the hugely expensive experiments being conducted at CERN, there was a need to store and present information in a way that overcame the incompatibilities between different computers. Email was a well established method of passing information from one computer to another, but it did not provide a public space in which information could be immediately presented to anyone who asked for it. Furthermore, email did not provide any mechanism for linking together related pieces of information to guide the user towards data that they were looking for.

Berners-Lee hit upon the idea of using hypertext to provide these linkages. Hypertext was not a new idea. The index used in a printed book can be considered to be a form of hypertext and hypertext had already been used to provide cross-references within electronic documents. However, Berners-Lee was the first person to apply the idea to a computer network. Using hypertext, it would be possible to link text in one document to related text in another document—irrespective of where the two documents were stored on the network. As Michael Dertouzos, the Director of the MIT Computer Science Laboratory, subsequently commented:

> "Thousands of computer scientists had been staring for two decades at the same two things—hypertext and computer networks. But only Tim conceived of how to put those two elements together to create the Web."[4]

Berners-Lee was determined to keep his design as distributed as possible. He did not want each web page or hypertext link to be registered in a central database—that would restrict the growth of the web and might enable governments or powerful corporations to control freedom of speech on the new medium. People should be able to create and modify their own web pages without having to submit them for approval. Establishing links between one web page and another should be trivial.

To implement this idea, a new form of document address would be needed. It would not only have to specify a file name for the required document and the directory in which that file was stored—it would also have to specify the particular computer on which that directory was located. To meet this requirement, Berners-Lee defined a "Uniform Resource Locator" (URL). A URL is the address that you type into your browser when you want to view a web page. A simple example is:

*http://www.google.co.uk*

Starting from the left, the http:// indicates that this is an Internet resource that is accessible using the HyperText Transfer Protocol (HTTP)—rather than some other protocol, such as FTP. HTTP is a standard protocol for transferring a web page from the computer on which it is stored (known as a "web server") to the browser through which it is viewed. The ":// " characters are used to separate two different parts of the address. They are followed by the website domain name (www.google.co.uk), which allows the browser to identify the web server on which the required information is stored. With this information, it is possible to retrieve the required web page.

A URL such as *http://www.google.co.uk* points to the main entry point (or Home Page) for a website. By convention, the home page is stored in a file called index.html. We could therefore have achieved the same result using the URL:

*http://www.google.co.uk/index.html*

Of course, there is usually more than one page associated with a website and the pages are often organized hierarchically—rather like the folders in Microsoft's Windows operating system. If we wish to view something other than the home page, the URL might look something like this:

*http://www.google.co.uk/intl/en/about.html*

In this case, the required web page is stored in a file called "about.html". The "intl/en/" probably indicates that it is stored in the international section of the website, in the English-language sub-section.

The ".html" file extension at the end of the URL above indicates that the file contains HyperText Markup Language (HTML). This is no different from the use of ".doc" to indicate a Microsoft Word file or ".pdf" to indicate an Adobe Acrobat file. HTML is a text-based computer language that can be used to define the layout and appearance of a web page. Fortunately, modern web authoring tools have largely eliminated the need for developers to write directly in HTML, but your browser should be able to display it if you want to see what it looks like.[5]

When you type in the URL for a web page that you want to view, your browser has to convert the URL into an IP address (we will explain how it does this when we discuss the Domain Name System). This IP address identifies the web server on which the web page is stored. The browser then sends an HTTP request to the web server to obtain the file containing the web page. When the file is received, the browser reads the HTML text and builds up the web page that the user sees on the screen. In some cases, the browser finds that images or other objects have been embedded in the web page and it has to make further requests to retrieve these additional files before the page can be displayed correctly.

In March 1989, Berners-Lee submitted a proposal to CERN to develop a hypertext-based information management system. After receiving no response, he resubmitted the proposal in May 1990. Although this, too, generated no formal response, Berners-Lee's remit at CERN was sufficiently flexible to enable him to begin work on a prototype. By mid November, he had created a browser/editor called WorldWideWeb that could decode URLs and display the contents of an HTML file. He had also created the software for the world's first web server, which held the specifications for the three key building blocks of the web (HTTP, URL and HTML) along with project-related information.

This prototype system was sufficient to demonstrate the principles of the World Wide Web, but it was far from being in a state in which it could be rolled out to ordinary users. For example, the software would only run on one—rather unusual—type of computer and the WorldWideWeb browser proved to be too sophisticated for many of the computer users at CERN. Berners-Lee responded to these problems by producing a line mode browser that could run on any computer. During 1991, web servers started to appear in other European and American institutions. By the summer of 1991, the CERN server was experiencing about 100 hits (requests for web pages) per day. One year later, this rate had increased to 1,000 per day and a year after that, it had reached 10,000 per day. In April 1993, CERN finally decided to release the WorldWideWeb software and protocols to the world—free, gratis and for nothing. This magnificent gesture was a key factor in the subsequent success of the web.

At about the same time, the National Centre for Supercomputing Applications in Chicago released the first version of the Mosaic browser. This software could be downloaded over the Internet and it made the web available to users of PCs and Apple Macintoshes via a simple point-and-click interface. Within a few months, a million copies of the software had been downloaded. It wasn't just scientists who were using the web to publish their research—the world was starting to realize that the web was a very efficient way of making all kinds of data publicly available. Rather than responding individually to requests for information, an organization could simply publish that information on the web and let people help themselves.

The software developers behind Mosaic subsequently set up Netscape to commercialize their development. After a dispute over the Mosaic name, the browser was re-launched in April 1995 as Netscape Navigator 1.0. The software was fast, stable and feature-rich, and soon became the de facto standard for web browsing. The extraordinary success of Netscape Navigator demanded a competitive response from Microsoft, who had been slow to recognize the true significance of the Internet. In August 1995, they released Internet Explorer 1.0. It was initially made available as part of an add-on pack for Microsoft's Windows 95 operating system. However, the tighter integration with Windows that appeared in subsequent releases led to accusations of unfair competition and resulted in well publicized legal action. Naturally, the intense rivalry between Netscape and Microsoft (often referred to as the "Browser Wars") led to rapid improvements in browser technology and helped to drive the phenomenal growth of the web.

The development of the World Wide Web has dramatically enhanced the capabilities of the Internet. It has led to fundamental changes in the way that companies do business and has become an indispensable part of modern life. It provides access to a mind-boggling amount of information covering every imaginable field of human endeavor. Many of the applications described later in this chapter are built upon foundations provided by the World Wide Web.

## 10.3 Domain Name System (DNS)

A packet of data cannot be sent across an IP network unless it has a valid destination address. An IP address looks something like this:

123.45.6.78

A string of numbers separated by dots may be exactly what a router likes to see but it is rather less satisfactory for the average human being—most people find it simpler to memorize a hostname such as *www.google.co.uk*. If we wish to specify network destinations using names rather than numbers, then we need some way of converting from one to the other.

In the early days of the ARPANET, this was done by the hosts.txt file. This file listed the names and the corresponding addresses of all the host computers connected to the network, rather in the way that a telephone directory lists the names and telephone numbers of users on a telephone network. The hosts.txt file was maintained by the Network Information Center (NIC) at Menlo Park, California, and each computer on the network would download the latest version every day. This simple solution worked perfectly adequately when a few hundred computers were connected to the network, but became increasingly unmanageable as the network started to grow rapidly. A system that required every address on the network to be managed from a single location was clearly not going to be scaleable! Furthermore, choosing hostnames that had not already been used was becoming increasingly difficult—particularly as a high proportion of new network users wanted their computer to be called Frodo, after the heroic hobbit from *The Lord of the Rings*.[6]

In 1983, Jon Postel, Paul Mockapetris and Craig Partridge proposed a hierarchical naming structure that would overcome these problems. After a considerable amount of debate (Internet people love a good argument!), the following top-level domains were created:

"edu"   university or other educational organization
"com"   commercial organization
"gov"   government
"mil"   military
"net"   network service provider
"org"   non-profit organization

Since then, additional top-level domains such as "biz" and "tv" have been defined for use by other types of organization, along with country-specific domains such as "uk", "fr" (France) and "jp" (Japan). A selection of top-level domains can be seen immediately below the root of the hierarchy in Figure 46.

Each of these top-level domains contains a number of subsidiary domains. For example, the hostnames "google.com", "yahoo.com" and "amazon.com" all exist within the "com" domain. Similarly, the "uk" domain subdivides into "ac.uk", "co.uk", "gov.uk" and many others, while "co.uk" subdivides into "virgin.co.uk", "bbc.co.uk" and so on. The first (left-hand) section of the address is specific to an individual computer and the address becomes increasingly general with each succeeding section. It is just like a postal address, which might start with a specific house name and finish with a country. Hostnames used in one domain can be re-used in another, so abc.co.uk is completely different from

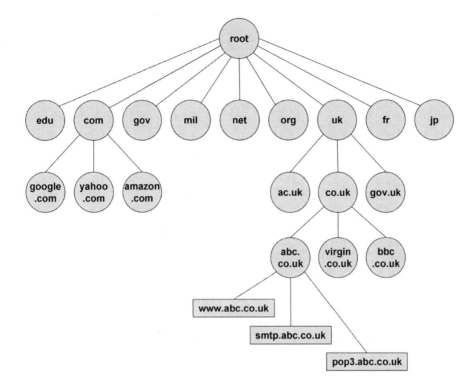

**Figure 46.** Domain name structure.

abc.com or abc.org.[7] Clearly, the diagram can only show a tiny fraction of the addresses that have been defined.

Domain names and hostnames may appear similar, but they are not quite the same thing. All hostnames have an IP address associated with them,[8] but this is not always the case with domain names. Some domain names (such as "co.uk") are needed to build the logical hierarchy, but do not provide the hostname of a physical computer and so do not have an IP address associated with them. In Figure 46, hostnames are shown in rectangular boxes.

A hierarchical address structure avoids the need for centralized address management, thereby overcoming the principal weakness of the hosts.txt system. For example, company ABC would have to register "abc.co.uk" with Nominet (the body responsible for allocating names within the "co.uk" domain). Once it has done this, ABC would then be free to create sub-domains of its own (or sub-sub-domains, or sub-sub-sub-domains) without having to ask anyone's permission. By convention, *www.abc.co.uk* would be the hostname of the computer hosting ABC's website, while the server handling their outgoing email might be called smtp.abc.co.uk and the incoming email server might be called pop3.abc.co.uk.[9] Notice that there is no possibility that ABC's sub-domains will conflict with anyone else's, because nobody else will have a subdomain ending in "abc.co.uk". Using a hierarchical address structure means that there is no need for all names to be allocated centrally—this responsibility can be delegated to the appropriate level in the hierarchy.

The hierarchical address structure provides an elegant solution to the problem of managing huge numbers of hostnames but we still need some way of converting from a hostname to the numerical IP address of the computer.[10] This problem was solved in 1983 with the invention of the Domain Name System (DNS) by Paul Mockapetris. As in the case of the addresses themselves, the DNS has a hierarchical structure, allowing information relating to network addresses to be distributed across hundreds of thousands of "name servers" in the network.

So, how does it work? Let's start by describing how it works in theory before considering some of the shortcuts that are used to speed things up in practice. In order to determine the IP address corresponding to the hostname *www.abc.co.uk*, your computer will need to interrogate a name server. Your computer will have been pre-configured with the address of the name server that sits at the top of the DNS hierarchy.[11] This name server will not know the IP address of *www.abc.-co.uk*—it cannot be expected to know everything—but it will return the IP address of another name server that handles addresses ending in ".uk". Your computer will then send the same request to this new name server, which, once again, will not know the answer. This time, however, your computer will receive the IP address of the name server that handles addresses ending in ".co.uk". Your computer will then interrogate this third name server and will receive the IP address of the name server that handles addresses ending in abc.co.uk. On the fourth attempt, your computer will finally receive the IP address of *www.abc.-co.uk*.

This process is beautifully logical, but could become rather slow and cumbersome if a complex hostname (such as *"whatson.herts.localnews.abc.co.uk"*) needs to be translated. Furthermore, the initial name server represents a potential bottleneck in the system, because it has to be accessed for every hostname translation on the whole Internet! These problems can be overcome using a technique called "caching" (pronounced *"cashing"*). Caching is based on the idea that users tend to access the same hostnames over and over again. If the result of each address translation is stored, then the cumbersome process described above is only required once for each address.

Caching can, in fact, be applied at multiple levels within the DNS hierarchy. For example, your web browser may cache recent address translations in case you decide to return to a page that you viewed earlier. Furthermore, if you attempt to access a popular website, the IP address may already be held in your ISP's DNS cache as the result of a previous access by another user. However, this raises another problem: how can you be sure that data stored in a cache is still valid? If the IP address corresponding to a particular hostname has been changed recently, how does that change make its way into the cache? The answer is that each hostname in a cache has a defined Time to Live; when that time expires, the hostname is deleted from the cache. The next access to that hostname causes the long-winded address translation method to be used and the new data is stored in the cache. Using this mechanism, all data held in the cache is periodically updated.

Each domain name on the Internet must be unique and the extraordinary growth of commercial activities on the Internet has made some domain names into very valuable assets. This has given rise to a phenomenon called "cyber squatting". Since a company's website is usually an important tool for sales and promotional activities, it is essential that customers can find their way to it without difficulty. Many website addresses can be constructed by simply placing "www" before the company's name and ".com" after it, so it is vital for the company to own this domain name. If, however, this address has already been registered by someone else, then gaining control of it can be an expensive business.

In the early days of the Internet, many companies were slow to register a suitable domain name. When they finally realized the significance of what was happening, they often found that their preferred domain name had been registered by a speculator hoping to make a quick buck. In some cases, the speculator increased the pressure on the company to reach a financial settlement by posting unflattering comments about the company on a website associated with the domain name. Processes now exist for resolving domain name disputes, but start-up companies are finding it increasingly difficult to find memorable domain names that have not already been taken. Even if a particular name is available in one domain (such as "co.uk"), it may not be available in other countries in which the company plans to operate. In a number of cases, expensive re-branding exercises have been needed because the required domain names were not obtained early enough.

The DNS is such a fundamental part of most people's Internet usage that it is often seen as an integral part of the Internet rather than as an application running on top of it. Popular applications such as email and the World Wide Web make use of the facilities provided by the DNS, but it is perfectly possible to use the Internet without using the DNS and a number of alternative address resolution technologies have been developed to meet the needs of specialist applications.

## 10.4 Search engines

Imagine trying to find a book in a library where the indexing system has been lost and where some hooligan has dumped all the books in a pile on the floor. You might be able to find a particular book if someone was around who knew where it was but you would have very little chance of finding it on your own. Now let's imagine that you need to be able to find a particular word or phrase within a particular book. Just to make things a bit more interesting, let's also imagine that the number of books in the library is measured in billions rather than thousands and that new books are appearing every second. It's a problem of such mind-blowing complexity that it is amazing that technology can solve it at all. Yet, that is effectively the problem that Internet search engines solve every day with apparent ease. Search engines convert the Internet from an anarchic mess into a valuable tool. As a result, control of a popular search engine can confer huge economic power.

Internet search engines have been around since the early 1990s. However, early search engines—while better than nothing—could still be pretty frustrating to use. Often, the information that you were looking for was buried in page after page of unimportant or completely irrelevant results. Web pages normally have keywords associated with them and the search engines simply looked for keywords that matched the user's search criteria. A search for "BBC" might throw up thousands of pages that mentioned the BBC, but the BBC's main website might not appear anywhere near the top of the list. This approach generated large numbers of low-grade results and could be manipulated by unscrupulous website owners to increase the prominence of their own websites.

Today, Google is the undisputed leader in Internet search engines. It has achieved this position in spite of not launching until 1998—by which time a number of other search engines were already well established. One of the factors that has set Google apart from its competitors is the quality of its search results. Rather than simply looking for keyword matches, Google seeks to assign a ranking to every web page that it searches. The exact details of Google's PageRank algorithm remain a closely guarded secret, but some of the basic principles are known.[12] For example, Google assigns a higher value to a web page if other websites have links to it. Furthermore, those links are worth more if they come from a page that has already been assigned a high value, so a link from the Yahoo! home page would be worth far more than a link from the Wheen family's holiday photos. When searching for multiple words, Google gives priority to those pages in which the words are close to each other in the text. It even looks at the font size, because words in larger fonts are likely to be titles or section headings. Clearly, this approach needs some serious computer power, but Google has broken the process down so that each part can run on a separate computer. In the early days of the company, the search engine ran on a large number of standard PCs. As its popularity grew, they simply bought more PCs.

Google uses a "web crawler" to find and catalog websites. A crawler is a piece of software that wanders around the web following links from page to page. Each new page that it finds is analyzed and the results are stored for use in future searches. Pages that have been visited previously are re-checked periodically to see whether they have changed (some pages, such as news sites, are likely to change very regularly). Although search engines tend to be used to search for web pages, they can also be used within corporate intranets and search software can be downloaded to a PC to find information stored on the hard disk. Specialist search engines now exist for finding particular types of information such as images, academic papers, blogs, discussion groups, chat rooms and databases.

Google is a hugely profitable company, but its searches are provided free of charge and its home page carries no advertising. Google makes most of its money from the "sponsored links" that appear above and alongside the search results.

Unlike conventional advertising—where most of the recipients have absolutely no interest in the product being promoted—Google delivers advertisements that are specifically targeted at the recipient. If someone is searching for "Used Cars", Google will supply sponsored links to car dealerships. Searching for "Investment" will produce sponsored links to providers of financial services. Google works much more effectively than traditional advertising because it delivers highly targeted information to users at precisely the moment at which they are looking for it.

The cost of advertising on Google depends upon the product or service that is being promoted. Clearly, every advertiser would like their sponsored link to be at the top of the first page of search results, so the more prominent sponsored links are more expensive. Similarly, some products and services are likely to generate more revenue per customer than others, so the value of the sponsored links will vary accordingly. Google manages this situation by holding continuous online auctions for the sponsored links associated with particular keywords. Companies pay for each click on one of their sponsored links and the company bidding the highest amount per click gets the best slot on the page. Clearly, managing an advertising budget on Google can require specialist expertise and search engine marketing consultants are available to help—at a price, of course!

## 10.5 Bulletin boards, newsgroups, discussion groups, blogs and wikis

In the early days of the ARPANET, email discussion groups enabled groups of users to hold online discussions about subjects of mutual interest. The concept was pretty simple. If you wanted to participate in a discussion, you sent an email to the group administrator asking to join. Assuming that you were accepted, your email address would be added to the distribution list. You would then be able to send email to the other members of the group and you would also receive contributions made by other members.

Discussion groups initially focused on technical issues, but the scope soon broadened out to include science fiction and other subjects that ARPANET users felt strongly about. Since the discussions were taking place via email, all contributions to a discussion had to be downloaded to the user's computer before they could be read. This was a potential problem. Although some contributions might be immediately deleted, participation in a number of discussion groups could generate a significant amount of email and computers at that time had very limited storage capacity. Furthermore, you needed to join a discussion group before you could see whether their discussions were of any interest and you then had to leave them again if you wanted the emails to stop—there was no simple way of browsing through discussion groups looking for items of interest. Finally, discussion groups were restricted to the privileged few who had access to an email account on the ARPANET. Discussion groups might be simple, but there was scope for improvement.

In 1977, the first public bulletin board appeared. A bulletin board was the electronic equivalent of a kitchen noticeboard—it could display useful or interesting information (newspaper articles, bus timetables, family photographs, etc.) and could be used to leave messages. Early bulletin boards were set up by enthusiasts using their own software running on home computers such as the Apple II and the Radio Shack TRS-80. Users would dial into the bulletin board using modems running at 110 or 300 Baud, so performance was painfully slow. Bulletin boards did not provide the networking capabilities of the ARPANET,[13] but they did enable software and data to be downloaded and they did provide users who had no access to the ARPANET with a forum for online discussions.

Bulletin boards tended to attract users who lived locally, because people who lived further away were put off by the long-distance telephone charges for the dial-up connection. The owners of bulletin boards turned this restriction to their advantage by providing coverage of local issues and by organizing events at which bulletin board users could meet. By the early 1990s, bulletin board services had diversified to include puzzles, games and video clips. Bulletin board users could exchange email with each other and some bulletin boards provided gateways to allow Internet email to be sent and received. These services were extremely popular and some bulletin boards started to generate significant revenues. However, the arrival of the World Wide Web in the mid 1990s spelled the end for the bulletin board. A few (such as AOL) developed an Internet presence and evolved into ISPs, but the stand-alone bulletin board with dial-up access disappeared into history.

In 1979 (shortly after the appearance of the first public bulletin board), a computer at Duke University in North Carolina was linked to another computer at the University of North Carolina. Since an ARPANET connection was not available, a 1,200 bit/second dial-up modem link was used instead. Both of the machines were running the Unix operating system and Unix had just acquired a facility that would search for changes to specified files on one machine and copy them across to the other. If a file was created on one machine to hold (say) details of all the latest software bug fixes, then each update to the file would be propagated across to the other computer whenever it dialed up a connection. With a bit of simple programming, it was possible to automate this dial-up process so that people viewing the copy of the file could be sure that it was reasonably up-to-date. Of course, there was no reason why this file synchronization mechanism should be restricted to just two computers—other computers in different parts of the country could also dial in and obtain their own updated copy of the file.

It was soon realized that this mechanism could be used to distribute a much wider range of information. The software was enhanced, new computers were connected and Usenet was born. Before long, the system was distributing games, software and email, but it was the Usenet discussion groups (known as "newsgroups") that really caught the public imagination.[14] The extraordinary proliferation of Usenet newsgroups on different subjects led to the definition of a hierarchical structure to help people locate particular groups that they were interested in.[15]

Of course, anyone viewing a copy of a file (rather than the original) could never be quite sure that they were getting up-to-the-minute information—there was always the possibility that the master file had been changed since the files were last synchronized. However, if a dial-up link was established once or twice a day, the users could be sure that they were seeing something that was close to the latest version. Usenet might not provide the full facilities of the ARPANET but it had the big advantage that it was available to just about anyone with access to a computer, a modem and some suitable software. Furthermore, Usenet was organized differently from bulletin boards and access was not constrained to a particular geographical location by long-distance call charges. It was, in effect, a "poor man's ARPANET".

As Usenet and ARPANET expanded, it was inevitable that they would eventually meet. The first bridge between the two systems was created at the University of California and it meant that ARPANET discussions became available through Usenet. Special "newsreader" software was required to access Usenet but Tim Berners-Lee—recognizing the importance of Usenet—incorporated newsreader software into his first web browser. As a result, a huge amount of existing information became available on the web and this helped to drive the adoption of web technology. Today, bulletin boards and Usenet newsgroups have evolved into Internet forums that are accessible using a standard web browser and Usenet archives dating back as far as 1981 are available on the web via Google Groups.

In recent years, there has been considerable interest in Internet blogs. A blog (short for "web log") is essentially an Internet forum that is used by a single individual to record their thoughts and opinions. Some provide commentaries on particular topics, while others are used as a form of online diary. A typical blog will contain a mixture of text, pictures and links to other web pages. Facilities are normally provided for readers to comment on the blog but, unlike a newsgroup or discussion group, this is not a conversation between equals; the essence of a blog is that it provides one individual's (or one organization's) view of the world.

There is now a huge number of blogs available on the Internet and the "blogosphere" is growing all the time. Some—such as the Baghdad blogger during the Second Gulf War—become popular because of the personal insights that they provide into important events. Other popular blogs deal with much more mundane issues, but do so in an interesting or humorous way. At the other end of the scale, some blogs are little more than self-opinionated rants or platforms for publicity-seekers. Early blogs were often manually updated personal websites, but the development of blogging software has opened up blogging to people with less technical skill.

Wikis, like blogs, have evolved out of Internet forums. The name comes from the Wiki-Wiki Express shuttle bus that runs between terminals at Honolulu International Airport ("wiki" is the Hawaiian word for "fast"). A wiki is a web-based tool for mass collaborative authoring that allows visitors to add, remove or change content. In most cases, there is no requirement for new contributions to be reviewed before they are accepted and there is often no need for contributors to register. The online encyclopedia, Wikipedia, is a well known example of a wiki. As in the case of blogs, wikis have found applications within large organizations as a means of providing project communication and documentation.

Although wikis are a relatively recent innovation, they actually bring things closer to the original concept for the World Wide Web. Tim Berners-Lee was disappointed when the web came to be seen as:

> ". . . a medium in which a few published and most browsed. My vision was a system in which sharing what you knew or thought should be as easy as learning what someone else knew."[16]

## 10.6 Instant messaging and chat rooms

Instant messaging applications first appeared on mainframe computers in the late 1960s. Some multi-user operating systems provided facilities for messaging between users who were logged on to a particular machine and the concept subsequently spread to Local Area Networks. In November 1996, a company called Mirabilis launched free instant messaging software for use on the Internet and instant messaging really took off.

During an instant messaging session, a small window is opened on each computer so that the participants can see what has been typed so far. Although typically text-based, most popular instant messaging systems also allow sounds, photographs, files and web page links to be transmitted and an Internet-based telephony service is usually included as well. Unlike email or Internet forums, instant messaging requires all the participants to be online at the same time, so instant messaging systems normally keep track of a user's status (available/busy/offline, etc.). Users who do not wish to be disturbed can make themselves unavailable for instant messaging.

Instant messaging involves no significant time delays, so it is much closer to a genuine conversation than email. This can be useful in situations in which a response is needed immediately or in which a sequence of questions and answers is likely to be necessary. Whilst instant messaging is not as quick as making a phone call, some people consider it to be less intrusive because it does not force an immediate response.[17] Free instant messaging services are available on the Internet, so it is a form of communication that appeals to financially challenged teenagers and to travelers who wish to avoid the cost of international phone calls. Instant messaging also opens up new communication options for people with hearing or speech impediments.

Instant messaging conversations are not restricted to two people and multi-way chats are probably rather simpler to establish than telephone conference calls. Private "chat rooms" can be established by groups of friends and can also be used in a business context to facilitate communication between workers on a particular project. Participation in a private chat room is restricted to people named on a contact list, but public chat rooms are also available for online conversations with anyone who shares a particular interest.

## 10.7 Podcasting and RSS

Let's suppose that you are a fan of a particular blog, but it is not updated on a regular basis—sometimes, there might be three or four new postings in a single week, while on other occasions, there are no new postings at all for more than a month. You could make a point of checking the blog every day, but you would often find that no new material had been added.

Now let's suppose that you like to keep track of other blogs as well. You might also be an active participant in some Internet forums and perhaps you like to download the latest news every day before breakfast. Checking a large number of websites on a regular basis can be a rather tedious and time-consuming job. A better solution would be to install some software on your PC that periodically checks each of the sites on your behalf. Fortunately, this software is readily available on the Internet; it is called a web feed aggregator (or simply an aggregator) and it runs in the background whenever your PC is turned on. It regularly scans each of the sites that you have selected and downloads any new information that it finds.

The web feed aggregator uses a technology called RSS. As you surf the web, you may have noticed an icon (RSS logo) that looks like this:

RSS Logo 1

or possibly this:

RSS Logo 2

This indicates that an RSS feed is available if you want it.[18] A feed is simply a list of Internet addresses from which files can be downloaded and these files might contain (for example) successive episodes of a radio program. Clicking on this icon causes details of the feed to be added to the list of Internet sites that your aggregator will keep an eye on. Whenever new information becomes available, the aggregator will automatically download it to your PC.

This mechanism has created a new form of Internet broadcasting called "podcasting". The "pod" in the name is a reference to Apple's iPod device, because podcasts are often downloaded via the PC to some form of portable media player. As an example, the BBC morning news is available as a video podcast for viewing on the train while commuting to work. Podcasting is also used in the education field for communicating administrative information between colleges and their students and for distributing lecture notes and slides. A podcast would typically be an audio or video file, but the same mechanism can also be used to download text or any other form of content.

One of the first significant applications of this new form of broadcasting occurred in November 2002 when the *New York Times* began offering a range of RSS feeds relating to news topics. Since then, podcasting has been embraced by most mainstream content providers but the technology also opens up opportunities for distributing niche content to much smaller audiences. A conventional radio or TV station has substantial costs associated with transmitters and other infrastructure, but podcasting is so cheap that teenagers can use it to broadcast from their bedrooms! While the radio spectrum is very congested, there is no real limit on the number of separate podcasts that can be supported by the Internet.

## 10.8 Peer-to-peer networks

The World Wide Web uses a computing architecture known as "client–server". As you view a web page, the client is the web browser running on your laptop computer, while the server is the computer on which the website is stored. When you click on a hypertext link, the client will send a request to the server holding the required web page, the server will transmit the web page over the network and the client will display it on the laptop's screen. Of course, client–server architectures are not confined to the World Wide Web—they can be used for any application in which a client computer is used to request services from a server located somewhere else on a network.

In recent years, there has been considerable interest in the development of peer-to-peer computer architectures. As the name implies, a peer-to-peer network is a connection between equals, so, for example, a peer-to-peer relationship could be established between your laptop and a friend's computer. In contrast to the client–server model, each computer in a peer-to-peer architecture can behave as either a client or a server as the need arises. This key difference between peer-to-peer and client–server architectures is illustrated in Figure 47.

Peer-to-peer concepts have been around since the earliest days of the Internet, but it is only in recent years that they have gained widespread public attention. This has occurred mainly as a result of peer-to-peer file-sharing networks such as Napster. Napster was developed by Shawn Fanning in June 1999 while he was

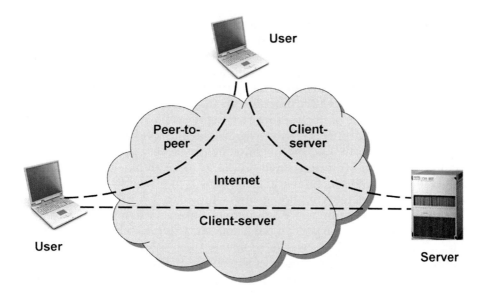

**Figure 47.** Peer-to-peer and client–server.

still a student at North Eastern University in Boston. The software was widely used for illegal swapping of digital music. The Recording Industry Association of America (RIAA) was quick to respond with a lawsuit and in March 2001, they won a temporary injunction against Napster to stop it from transmitting copyright music over its network. After bankruptcy and a buyout, Napster eventually re-launched in October 2003 offering legal music downloads to paying customers. Another high-profile peer-to-peer file sharing service, Kazaa, finally converted to a legitimate download service in July 2006 after paying $100m in damages to the recording industry.

The use of peer-to-peer networks for illegal file sharing has given the technology a controversial reputation, but it has many legitimate uses. For example, the BBC has used peer-to-peer networking as a cost-effective way of allowing people to catch up with TV and radio programs for up to 7 days after they were broadcast. A technology called Digital Rights Management (DRM) was used to ensure that the BBC retained control over the program material.[19]

The Skype Internet telephony service provides another high-profile example of peer-to-peer networking. An early version of Skype was released in August 2003 and by March 2007, the software had been downloaded more than 500 million times. This phenomenal rate of growth owes much to the fact that there is no charge for the software and calls between Skype users are also completely free. This has made Skype an attractive alternative to international phone calls— particularly as the calls can be routed to a Skype user irrespective of where they happen to be located. Although there is a charge for Skype calls that originate or terminate on a conventional telephone network, the charge is usually small.

The technical details of the Skype system are a closely guarded secret.

However, it is known that users log onto the network via a centralized login server, so this is not a pure peer-to-peer architecture. Furthermore, some of the more powerful computers running the Skype software are promoted to "super nodes" and are required to perform additional functions on behalf of other users. In spite of this, ordinary telephone calls between Skype users are genuinely peer-to-peer. The audio bandwidth provided for the call depends upon the available capacity in those parts of the Internet carrying the call—if there are plenty of bits available, Skype can offer roughly twice the audio bandwidth of an ordinary telephone call and call quality will consequently be very good.

Grid computing is often cited as another example of a peer-to-peer architecture.[20] The idea behind grid computing is to make use of spare processor cycles when a computer is not doing any useful work. By installing the necessary software on a large number of computers linked to a network, prodigious amounts of computing power can be assembled. This power can then be used to address computationally intensive problems such as weather forecasting and scientific data analysis. One application that has particularly caught the public imagination is the SETI@home project, which is using grid computing techniques to search for extra-terrestrial life. By August 2008, over five million computer users in more than 200 countries had downloaded the free software and the performance of SETI@home was rivaling the fastest supercomputers on the planet.[21]

These examples illustrate that peer-to-peer applications can scale to massive sizes without hitting capacity limitations. Furthermore, they can be highly resilient because additional resources are always available to replace failed parts of the network. Peer-to-peer applications have proliferated on the Internet in recent years and further innovation can be expected.

### 10.9 Virtual private networks

Many organizations operate private networks to link up their various offices around the country—or around the world. A private network can be constructed using fixed bit rate digital circuits leased from a network operator, but similar results can be achieved using the Internet. A virtual private network (VPN) makes use of the same physical infrastructure as any other Internet application, but users of the VPN are presented with the appearance that they have their own private network. Most of the benefits of a traditional private network are preserved within the VPN, but usually with lower costs and enhanced flexibility. Further details are provided in Appendix J.

### 10.10 Voice

The idea of transmitting voice signals over packet-based networks dates back to the very beginnings of packet switching. In the early 1960s, Paul Baran realized that his fault-tolerant network concept would have to be digital if it was going to carry voice over long distances because the store-and-forward operations occurring at each node would seriously degrade an analog signal.[22] Furthermore, digital voice can also be encrypted much more securely than analog voice, making it more suitable for military applications. The Network Voice Protocol was developed for the ARPANET in 1973 and experiments with digital voice continued in the years that followed. However, it was the success of ARPANET's data applications (such as email) that captured the headlines and it appears that digital voice was little more than a sideshow in those early days.

A significant step forward occurred in 1995 when an Israeli company called VocalTec released "Internet phone" software that enabled voice communication across the Internet. Voice calls were restricted to computers that were running the VocalTec software, users had to wear headphones and voice quality was generally poor but—hey—the calls were free! A computer may not bear an obvious resemblance to a telephone but it has a keypad for dialing, it can handle audio signals and call information can be displayed on the screen. For some applications, a PC-based "soft client" is still a convenient way of making phone calls. However, many people prefer the simplicity and convenience of a standard telephone. As dial-up Internet access was displaced by always-on broadband services, broadband interface devices were developed to connect standard telephones to the Internet.[23]

Internet telephony has become a serious threat to the public telephone network, offering cheap (often free) calls with a wide range of attractive service features. However, anyone who has used an Internet phone service will know that voice quality is still far from perfect when compared to the existing telephone network. Appendix K explains how telephone calls are carried over the Internet and discusses how some of the problems are being addressed.

## 10.11 Video

Having solved the problems of carrying voice across the Internet, it was only natural that people would start to look at ways of transmitting video. Video, like voice, can be converted from analog to digital and transmitted as a sequence of packets. However, video requires far more bandwidth than voice and, until recently, the Internet struggled to provide sufficient capacity. As a result, the quality of Internet video was often pretty dreadful.

To understand why video requires so much bandwidth, let's consider a typical computer display. The display is made up of an array of "pixels",[24] where a pixel is the smallest dot that the screen can display. The display on my laptop has a resolution of 1,400 × 1,050 pixels, so the image that I see is actually made up of 1,400 × 1,050 = 1,470,000 separate dots. Fortunately, each pixel is so small that I just see one continuous image. The digital video signal must also specify the color of each individual pixel. If 8 bits are used to specify the color of one pixel, then it would be possible to specify 256 ($2^8$) different colors. In practice, this produces rather unsatisfactory results, so my laptop uses 32 bits per pixel. If we multiply the number of bits per pixel by the number of pixels on the screen, we find that over 47 million bits are required to specify a single image on the display.

This may seem like a huge number, but things are about to get worse. A movie camera typically records about 25 separate images per second and it is by displaying each of these images for just 1/25 of a second that the illusion of motion is created. If we needed to transmit 25 images per second—with 47 million bits per image—then we would need a bandwidth of more than 1 Gbit/second. No wonder the Internet finds this hard to handle!

Of course, we can reduce the bandwidth required by reducing the size or the definition of the video picture (thereby reducing the number of pixels) and this approach is often used on websites such as YouTube. Furthermore, there are video compression techniques that can be used to reduce the number of bits. To illustrate how video compression works, consider a television news broadcast. If the newsreader is sitting relatively still, then an individual pixel somewhere in the top left-hand corner of the screen will probably not change at all from frame to frame because it is simply providing part of the studio background. If we transmit the differences between one frame and the next—rather than recreating each successive frame from scratch—it requires far fewer bits. Further bit rate reductions are possible if the newsreader is sitting in front of a plain background because many adjacent pixels at the edges of the screen are likely to be exactly the same color.

As with any form of digital content, video can be stored as a file on a computer. For very low-speed Internet connections, one way to deliver video is to download the file to the user's computer prior to viewing. It might take a minute to transfer a file holding only 10 seconds of video, but that is the price that you pay for having a low-speed Internet connection. On higher-speed connections, the transfer of video information can keep up with the rate at which it is being viewed, so "streaming" can be used. Streamed video is displayed almost as soon as it arrives at the receiver, with just a small amount of buffering being used to smooth out any unevenness in the rate at which packets are arriving. With streamed video, 10 seconds of video takes 10 seconds to deliver.

Transmitting low-quality video over the Internet is one thing, but transmitting television signals is something entirely different. The bandwidth required to stream standard-definition television is about 2 Mbit/second, even if modern compression technologies are used, and this bandwidth rises to about 8 Mbit/second for high-definition television. Most standard broadband services cannot deliver such high bandwidths on a continuous basis, but it is possible to deliver "IPTV" television services using high-speed broadband networks that have been optimized for this purpose. In fact, "triple play" services (television, telephony and high-speed Internet access) can now be delivered simultaneously down a single telephone line. The techniques required to do this are discussed in Appendix L.

## 10.12 Social networking

In recent years, there has been dramatic growth in the use of the World Wide Web for social networking. Whilst sites such as Facebook and Twitter have captured most of the headlines, they are by no means the only ones—there are sites that focus on particular aspects of social networking (Friends Reunited, Flickr), particular age groups (TheStudentCenter, Sconex), particular nationalities or ethnic groups (Hyves, MiGente, CyWorld) or particular interests (CarDomain, LibraryThing, Care2). These sites are mainly intended for private socializing, but there are also sites such as LinkedIn and Xing that provide networking for business and career development.

Facebook started at Harvard University in 2004 and its initial growth was largely confined to American college students. However, the site is now open to people of all ages and its coverage is truly international. Facebook hosts a large number of separate social networks based around schools, companies or regions. Each user can publish a personal profile on the site, providing such diverse information as education, religion, sexual orientation, hobbies, life experiences and recent activities. Profiles can include photographs, videos and links to web pages and users can share their thoughts and opinions by writing notes or by blogging. A personal news feed provides a summary of all the latest activities that Facebook friends have been reporting. However, there are obvious risks associated with releasing too much personal information on the Internet, so access to personal profiles is restricted to friends and members of the user's networks and additional restrictions can be applied to specific individuals.

Facebook has opened up its software interfaces to third-party developers and this is bringing in a wide range of new applications to enhance the site. During the next few years, Facebook could become a platform for online applications in the same way as Microsoft's Windows operating system has become the platform for most desktop applications. This, in turn, could lead Facebook into direct competition with Google in areas such as online business applications that compete with Microsoft Office. Clearly, social networking has a strategic significance that extends far beyond simply linking up with a few friends!

## 10.13 Virtual worlds

A virtual world is an online simulation of an imaginary environment in which each user is represented by an "avatar". The avatar may or may not represent the user's real-life appearance. Avatars can communicate with other avatars that happen to be in their vicinity and this communication can take place over longer distances if they shout. Avatars can move around to explore their environment and can travel in vehicles ranging from go-karts to hot-air balloons. In general, the laws of physics apply in a virtual world in much the same way as they do in the real world, although teleporting is often available as a means of transport and avatars are sometimes allowed to defy gravity.

In appearance, a virtual world resembles a video game, but the similarity can be misleading. In a true virtual world, there is no real concept of winning or losing, and financial wealth is often the only way of keeping score. Virtual worlds have been included in this book because they provide a convenient environment in which communication can take place and people can interact with each other. Although voice and instant messaging can be used within a virtual world, the experience is much richer than a conventional online conference call or a chat room.

The concept of online virtual worlds dates back at least as far as 1987 and an online representation of the city of Helsinki became the first three-dimensional virtual world when it appeared in 1996. Since then, many different virtual worlds have been created. They are often targeted at particular age groups (such as teenagers) or people with particular interests (such as science fiction).[25] Inevitably, some adult virtual worlds offer virtual sex between avatars—although it may be necessary to purchase some virtual genitalia first. In November 2008, it was reported that a couple had filed for (real-world) divorce after the husband had been caught committing virtual adultery.

Second Life is a virtual world that has been enthusiastically adopted by certain sections of the business community. IBM owns 12 islands in Second Life; some of these are open to the public, while others are used for internal meetings and conferences. Dell owns a virtual factory in Second Life, where customers can configure computers to their own particular requirements, and Sears has a virtual showroom in which customers can redesign their kitchen. Sun has gained considerable publicity by making important strategy announcements from its pavilion in Second Life, while Toyota has started releasing virtual cars in Second Life to see how users react to new designs.

As in the real world, Second Life users are allowed to buy and sell assets. The virtual world has its own currency (the Linden Dollar), which can be exchanged for real-world currencies in Second Life's LindeX currency exchange.[26] Significant sums of money have been made by selling clothes, accessories and even walking styles for fashion-conscious avatars. However, Second Life's first self-made millionaire was Ailin Graef, who made her (real) money by buying virtual real estate, sub-dividing it, developing it and selling it. When it comes to making money, Second Life is surprisingly like the real thing.

However, Second Life is not just about commerce. Many universities and colleges have started teaching in Second Life and the results so far suggest that students are significantly more engaged in the virtual world than they are in other forms of distance learning. Artistic and creative activities are also flourishing in the virtual world. Second Life has stimulated experimentation with new forms of visual art and work can be exhibited in a virtual representation of the Louvre. There are also live music performances, providing a new channel through which real-world bands can gain exposure.

Marketing organizations tend to assume that people's behavior in a virtual world is the same as in the real world. The following incident may, or may not, illustrate this. When the French far-right National Front party, led by Jean-Marie

Le Pen, established a "virtual HQ" on an island in Second Life, the building was soon surrounded by protesters representing various left-wing organizations. A virtual riot subsequently broke out and the area echoed to the sound of heavy-caliber gunfire as the confrontation turned violent. One enterprising protester even created a pig grenade attached to a flying saucer and launched it at the National Front headquarters. However, it turned out that the riot took place in a part of Second Life that was not "damage enabled", so the result of this attack was nothing more serious than a spectacular explosion.

As if that wasn't enough, it was reported in August 2007 that Islamic militants were using Second Life to wage virtual jihad. Car bombings have occurred in the virtual world and the Australian Broadcasting Corporation's base in Second Life has been subjected to a virtual nuclear attack. It appears that terrorists may be using Second Life to practice skills such as reconnaissance and surveillance and there is a very real concern that Linden Dollars may be providing a convenient currency for terrorists to transfer funds across international borders.

\* \* \*

Networks today are no longer restricted to a single service—they are expected to carry voice, data, video, text and many other forms of content. In fact, networks today are no longer restricted to conventional services at all—applications such as Second Life are about far more than simply shifting bits from one place to another! A modern network is like a very large computer, providing a platform on which sophisticated applications can be built. Since access to processing power and storage can be provided through the network, it is becoming increasingly hard for users to draw any distinction between computers and networks. This phenomenon was summed up in 1984 by John Gage's famous phrase "The network is the computer". Although this concept was dismissed at the time by the computer industry, the extraordinarily diverse range of Internet applications described in this chapter illustrates just how right he was.

# 11    The mobile revolution

In 1984, Motorola launched the world's first commercially available handheld cellular phone, the DynaTAC 8000X. Popularly known as "The Brick", it measured 300 × 44 × 89 mm (13" × 1.75" × 3.5") and weighed a hefty 785 g (2 lb).

Motorola DynaTAC 8000X.[1]

In spite of its measly battery capacity (8 hours of operation in standby mode) and a $3,995 price tag, it sold in considerable numbers. The mobile phone revolution had begun. During the next 20 years, the mobile evolved from a rich man's plaything to a basic necessity of life. In most Western countries, the number of mobiles now exceeds the number of people. By any standards, that is an extraordinarily successful new technology!

Although the mobile phone is a relatively modern invention, the exact point at which the first mobile phone call was made is a matter of some controversy. As we saw in Chapter 6, some people have claimed that this occurred in 1879—just 3 years after Alexander Graham Bell invented the telephone—when David Hughes walked down Great Portland Street in London listening to a series of clicks broadcast from his home. Certainly, Professor Hughes was using a battery-powered telephone and the signals that he was hearing were transmitted by radio. On the other hand, no speech was transmitted and the communication occurred in one direction only. Claims have also been made for Alexander Graham Bell's Photophone (also discussed in Chapter 6), which was invented a year later. The Photophone used a mirror that vibrated in response to speech waveforms to modulate a beam of light. At the receiver, a light-sensitive selenium cell was used to generate an electrical signal that could drive a

A. Wheen, *From Dot-Dash to Dot.com: How Modern Telecommunications Evolved from the Telegraph to the Internet*, Springer Praxis Books, DOI 10.1007/978-1-4419-6760-2_12, © Springer Science+Business Media, LLC 2011

telephone earpiece. Although this device was capable of transmitting speech without wires, the need to maintain optical alignment between the transmitter and the receiver meant that it was not strictly mobile.

Clearly, we cannot determine who made the first mobile phone call without first specifying what we mean by a mobile phone. Prior to the Motorola DynaTAC 8000X, mobile telephones were normally mounted in cars (or boats or trains) because they were too heavy to carry around. Car phones were in use as early as the 1920s but, by stretching definitions a little, it is possible to argue that the car telephone appeared at least 10 years earlier than this. Lars Magnus Ericsson, who founded the LM Ericsson corporation in 1876, retired from business in 1901. From about 1910 onwards, it appears that he and his wife Hilda liked to tour the countryside in their car. Since Lars Magnus hated being cut off from the rest of the world, he carried a telephone in the car so that he could make calls whenever they passed a telephone line. Using two long rods, he would hook their telephone across a pair of wires and listen to see if the line was in use (early telephone networks used "open wire" cabling, so the wires were not insulated). If the line was free, they would crank the handle of their telephone dynamo to contact the operator in the local exchange. If the line was busy, they simply moved the rods to another pair of wires until they found a pair that was not in use. Of course, this technique was not truly mobile—the car had to stop first! Furthermore, it was restricted to outgoing calls and its legality is not entirely clear.[2]

It appears that some of the earliest users of radio-based car phones were members of the Police Department in Detroit, Michigan. The department had been using cars in a limited way since 1910, but it was not until the arrival of prohibition in the 1920s that the car became an essential tool of law enforcement—foot patrols and policemen on bicycles were simply no match for gangsters in highly powered automobiles. However, dispatching police cars to crime scenes remained a problem, so, in 1921, the department began experimenting with radios. Although some formidable technical problems were encountered, the experiment was judged a success. By 1928, radio-dispatched police cars were a permanent feature of policing in Detroit.

Radio brought dramatic improvements to Detroit policing, but it was still far from perfect. For one thing, communication was only one-way—the dispatcher could talk to the police cars, but the cars could not respond without stopping at a phone box.[3] Another problem was the lack of radio channels. The Federal Radio Commission had issued the Detroit police with a provisional commercial radio license, so the department was required to broadcast "entertainment during regular hours, with police calls interspersed as required". At one point, an exasperated police commissioner asked whether the police were expected to play a violin solo before dispatching a police car to catch a criminal.

As the use of radio expanded, spectrum congestion became an increasingly pressing problem. If a particular radio frequency was in use, then nobody else in the vicinity could operate on the same frequency without causing interference. Furthermore, limitations in the technology available at the time meant that even when transmitters were operating on different frequencies, a wide frequency separation was needed to avoid interference. When the first commercial mobile telephone system in the United States was inaugurated in St Louis, Missouri, in June 1946, it was allocated six channels, with a spacing of 60 kHz between channels. However, it was soon found that this spacing was not enough to prevent interference and the system capacity had to be reduced to three channels to fix the problem.

It is important to appreciate just what a serious restriction this was. Early mobile telephone systems used one central antenna per city and all radio telephones in the city communicated with that antenna. Channel frequencies could not be re-used in different parts of the city because that would lead to interference. If the system in St Louis was restricted to three channels, then only three calls to radio telephones could take place in the whole city at any one time.[4] It is not hard to imagine how such a severe restriction could stunt the growth of a promising new technology.

And there was another problem. Since each radio telephone had to communicate via the central antenna, it had to be sufficiently powerful to reach that antenna from anywhere in the coverage area. In some cases, that meant that it had to be capable of transmitting over distances of more than 50 kilometers. Whilst it was no problem to construct a large central antenna with this range, it was a major challenge for the designers of battery-powered mobile units. It was even suggested that some early car phones could only make two calls—the second one would be to the garage to tell them that the car battery was flat!

The system in St Louis addressed this problem through the use of zoning, as illustrated in Figure 48. The mobile user in his car (labeled (1) in Figure 48) is talking to a friend (2) on the fixed network (there were no mobile-to-mobile calls in those days). The central antenna (3) is powerful enough to transmit directly to the car, but the car's battery-powered transmitter is not powerful enough to reach the central antenna, so it communicates with another receiving antenna that is closer (4). Including the central antenna itself, there are five receivers distributed around the city (3–7). All the receiving stations are listening out as the car moves around the city and their outputs are combined to produce the signal received from the car. If the car moves across the city during the course of a call, then one receiver may produce a weaker signal whilst another one produces a stronger signal—in effect, the system has transferred the call from one receiver to another so that the conversation is not interrupted.

By 1947, the basic concept of a cellular network was starting to take shape. Rather than having one central antenna serving a whole city, the city would be divided up into a large number of "cells" and each cell would have its own central antenna (or "base station"). All the users in a particular cell would communicate via the base station serving that cell. Since the cells would be relatively small, the power required from the mobile transmitter would be greatly reduced, thereby reducing the drain on the battery. Furthermore, because the use

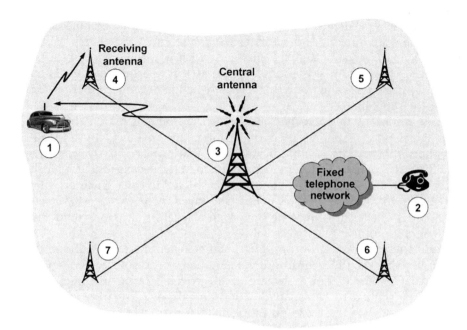

**Figure 48.**   St Louis mobile telephone system.

of lower-power transmitters meant that the signal did not carry very far beyond the boundary of the cell, the same frequencies could be re-used in other parts of the city without causing interference.[5] This important concept is illustrated in Figure 49.

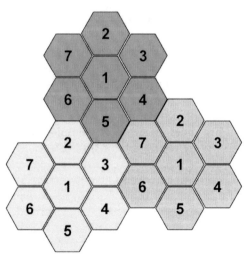

**Figure 49.**   Mobile network cells.

Three clusters of cells are shown, with seven cells in each cluster.[6] Cells marked with the same number operate using the same range of frequencies, but these cells are spaced sufficiently far apart that they do not cause interference to each other. Cells marked with different numbers operate using different frequency ranges, so they do not cause interference, even if they are physically adjacent. As the diagram shows, the seven frequency ranges used in one cluster of cells can be re-used over and over again in other clusters. As a result, the same radio frequency can carry different telephone calls in different parts of the same city and spectrum is used much more efficiently.

Of course, these advantages come at a price. For example, when a mobile user moved out of range of one receiver and into the coverage of another in the St Louis system, the mobile user could continue to operate on the same frequency. In a cellular network, any user roaming from one cell to another during the course of a call not only has to switch to a new base station, but also has to switch to a new frequency. Modern networks manage these complex "handoffs" without the user being aware that anything has happened, but the technology to implement this wizardry did not exist in the late 1940s.

During the years that followed, the lack of adequate radio spectrum meant that the growth of mobile networks was severely constrained in North America, while public radio telephones were not permitted at all in most other countries.[7] As a result, the technology developed rather slowly. By the mid 1960s, some mobile customers in North America could enjoy full duplex conversations (i.e. they no longer had to press a button to talk) and it finally became possible to dial calls directly without the need for operator assistance. Furthermore, there was no need to search the airwaves for an unused channel before making a call because this operation was now automated. On some networks, mobile-to-mobile communication became possible.

However, no true cellular networks had yet been built. The basic concepts— small cells, low transmitter powers, frequency re-use and inter-cell handoffs— had been known since the late 1940s, but early mobile radio networks used radio spectrum very inefficiently,[8] which meant that regulators were reluctant to support them by allocating additional frequency bands. Since it seemed certain that growth would continue to be constrained by spectrum limitations, there was little incentive to invest in new mobile radio technologies—in spite of the huge public demand. As a result, more than 20 years would elapse before the first true cellular network was built. In January 1969, a cellular network enabled passengers on the Metroliner train service between New York City and Washington, DC, to make telephone calls from payphones while racing along at more than 100 miles an hour. This impressive demonstration helped to persuade the Federal Communications Commission that the time had come to allocate additional radio spectrum to this important technology. Even so, it was not until the 1980s that the FCC finally allocated enough spectrum to enable mobile networks to demonstrate their full potential.[9,10]

At this point, we can revisit the question of who made the first mobile phone call. According to some accounts, this happened on April 3, 1973 when Martin

Cooper, John Mitchell and others demonstrated a portable cellular phone at a Motorola media event in New York City. According to the *New York Times*, "Mr Cooper, who placed the first call to a telephone within the room, dialed the wrong number and, after an embarrassed pause, said, 'Our new phone can't eliminate that'". The handset was a prototype and another 11 years would elapse before the first portable handset (Motorola's DynaTAC 8000X) became commercially available.

Whilst this might have been the first *publicly demonstrated* mobile phone call, it clearly wasn't the first call—the phone had been extensively tested before it was demonstrated! In view of the developments described earlier in this chapter, some readers may question whether this was really such a significant milestone. As we have said before, it all comes down to definitions. It had been possible to buy a radio telephone in a briefcase since about 1970, but it would probably be more accurate to describe these phones as "luggable" rather than portable and they were not designed for use on cellular networks. Motorola were able to demonstrate that their mobile phone was sufficiently portable to be used while walking the streets of New York City. If the essence of a mobile phone call is that the caller can be truly mobile (and not confined to a car), then this was an important moment.

In May 1978, the Bahrain Telephone Company began operating a cellular network with two cells and 250 customers. A few months later, AT&T conducted cellular network market trials in Chicago with over 2,000 customers.[11] In December 1979, a cellular network began operating in Tokyo. By the early 1980s, cellular networks had sprung up in Mexico, Saudi Arabia, Sweden, Norway, Denmark and Finland, and they spread like a rash across Western Europe. In the United Kingdom, the first two cellular networks launched service in 1985.

However, this rapid development of cellular networks created new problems. Many countries chose to develop their own cellular standards, so that by 1990, there were six incompatible standards within Europe and many others around the world. This meant that equipment manufacturers could not achieve real economies of scale, meaning that equipment prices remained unnecessarily high. Furthermore, a user from one country could not roam across an international border and continue to use their mobile in another country. These problems were seen as major impediments to the closer integration of Europe. In 1982, the first meeting took place to discuss a new cellular standard for Europe and 5 years later, the EEC committed itself to the GSM standard.[12]

The new standard had to meet certain objectives. In addition to driving down costs and enabling international roaming, it also had to support the use of portable handsets (not just car phones) and achieve good speech quality. Other objectives included support for a range of new services, higher spectral efficiency and compatibility with existing fixed network services. It was decided at an early stage that the best way to achieve these objectives would be to adopt an all-digital system and to abandon any attempt at backwards compatibility with existing analog mobile systems. Although this posed some challenges for the technology available at the time, the developers were confident that this problem would be resolved by further advances in speech compression and digital signal processing.

The first GSM services with international roaming were launched in 1991. By 2006, there were GSM networks operating in 214 countries and territories around the world and the total number of GSM subscribers exceeded 2 billion. In spite of being the product of a bureaucratic pan-European standardization effort, GSM has been stunningly successful. It is positive proof—if any were needed—that standardization of telecommunications equipment can deliver massive benefits.[13]

A description of the key parts of a GSM network is provided in Appendix M.

\* \* \*

The decision to make GSM an all-digital system meant that the handset had to contain a digital voice codec.[14] However, the limited radio spectrum available for mobile phone networks meant that it was impractical to encode each voice channel at 64 Kbit/second (as on the fixed network); speech compression techniques would have to be used to squeeze down the bit rate, even if this meant some sacrifices in speech quality. Early GSM networks used a speech coding technique[15] that produced adequate—but not wonderful—speech quality at 13 Kbit/second. Since then, more advanced speech coding techniques have been introduced,[16] but mobile networks still struggle to match the speech quality of the fixed network.

When you turn your mobile on, it establishes communication with a nearby base transceiver station (BTS).[17] Mobile base stations are a common sight these days—particularly near major roads and on areas of high ground—because large numbers of cells are needed if radio spectrum is to be re-used efficiently. The antennas that support the radio links are typically mounted at the top of a mast, while the BTS electronics are housed in a small cabin or building near the base of the mast. In urban areas, BTS antennas are often seen at the top of tall buildings, while in rural areas, the masts are some-

times disguised as (rather unconvincing) trees to reduce their visual impact on the landscape.

There is, in fact, a hierarchy of BTSs to meet different requirements. The largest are known as macrocells and they provide mobile coverage over a wide area. Then come microcells, which have a much smaller coverage area and are typically used in busy areas such as shopping malls and railway stations. Finally, there are picocells and femtocells, which have even smaller coverage and might be used to enhance mobile reception within a building.[18] Some femtocells are not much bigger than the average paperback book.

The area covered by a cell is determined by a number of factors, such as transmitter power, frequency and the height of the antenna. In theory, a GSM macrocell can be more than 30 kilometers in radius (although there are generally good reasons to keep cells much smaller than this), while a femtocell might be as small as 30 meters in radius. Clearly, a mobile phone needs to transmit considerably more power if it is linked to a base station 30 kilometers away than if it is linked to a femtocell in the same room. If the mobile phone transmits too much power, it will drain its battery unnecessarily quickly and may cause interference in adjacent cells; equally, if it transmits too little power, the radio link will not work at all. For these reasons, the power output from a mobile phone is constantly being adjusted by the network to ensure that it remains within acceptable limits.

When a cellular network is first built, the number of cells will be relatively small, so the radio planners will be primarily concerned with improving geographical coverage. They will tend to focus on macrocells, because these allow coverage to be established over a wide area using a small number of base stations. However, there will be gaps in the coverage where tall buildings or other obstructions block the path of the radio waves. If one of these gaps happens to be in an area in which people are likely to want to use their mobile phones, then a microcell might be needed to fill in the gap.

As the cellular network develops, significant coverage gaps will gradually be filled in. However, just when the network planners are starting to think that they have cracked the coverage problem, they will be hit by a new one: capacity. A new network is unlikely to have many subscribers, so traffic levels will be low. However, as the network grows, the number of subscribers is likely to increase and capacity bottlenecks will start to appear. Typically, these bottlenecks occur in public areas like railway stations, where mobile phone usage is likely to be high. Fortunately, microcells and picocells offer a way of increasing network capacity in these hotspots. Since they have very low power, these small cells can get away with using radio frequencies that would otherwise cause interference.

\* \* \*

Mobile networks use radio transmission, so eavesdropping is always a concern. While not perfect, the security features built into GSM networks represent a considerable improvement on the analog cellular networks that preceded them. A classic illustration of the vulnerability of analog cellular networks occurred in the early 1990s,[19] when a compromising conversation between Diana, Princess of Wales, and her lover, James Gilbey, was apparently recorded "off air" by two radio enthusiasts. Not surprisingly, these recordings eventually found their way into the tabloid press and the "Squidgygate" affair developed into another episode of royal soap opera. Although expert investigations subsequently cast doubt on whether the recordings had really been made by listening in to the airwaves, nobody denied that such interception was theoretically possible. Indeed, the "Camillagate" tape, which appeared at about the same time, does seem to be an "off-air" recording of a sexually explicit conversation between Prince Charles and his mistress—later, his wife—Camilla Parker-Bowles.[20]

With the introduction of digital GSM networks, it became possible to encrypt conversations to protect them from casual eavesdroppers, but security weaknesses remained. In 2006, it emerged that mobile phones belonging to the Greek Prime Minister, cabinet ministers and senior officials had been tapped for a period of about a year at the time of the Athens Olympics.[21] In Britain, the *News of the World*'s royal correspondent was jailed after hacking into a mobile network voicemail system used by Prince William.

A mobile phone has to be capable of changing to a different transmission frequency as it moves from one cell to another and GSM can exploit this capability to provide additional protection against casual eavesdropping. Rather than transmitting on just one frequency, "frequency hopping" systems jump rapidly from one frequency to another. The sequence of carrier frequencies to be used is continuously broadcast by the base station to each mobile phone operating in its cell. Anyone eavesdropping on one of the frequencies would just receive garbage, while a receiver that hopped between frequencies in step with the transmitter would hear the full transmission.

The origins of this idea are rather surprising. In 1942, the Hollywood actress, Hedy Lamarr, and the composer, George Antheil, were awarded a US patent for their "secret communication system". The system had nothing to do with movies or music (or mobile phones)—it was intended to prevent radio-guided torpedoes from being jammed by enemy action. Of course, the trick with frequency hopping systems is to keep the transmitter and receiver locked in step and Lamarr and Antheil's patent proposed something resembling a perforated piano roll for this purpose. This proved to be rather impractical and the patent expired before electronic technology had advanced sufficiently to enable the idea to be implemented.

Although frequency hopping can provide protection against eavesdropping and jamming, its primary role in GSM networks is to provide users with more consistent performance. If there is interference or poor radio propagation at a particular frequency, then frequency hopping will average out the problem across all the users in the cell. In effect, everyone's performance suffers a little bit—but not enough to matter. Without frequency hopping, a few users would suffer serious interference.

* * *

Up to this point, we have only considered the ability of mobile networks to carry voice. Although voice services continue to provide a substantial part of network operator revenues, data services are also very important. The time has now come to consider how mobile networks carry data.

We have seen how data access on fixed networks evolved from dial-up modem links to always-on broadband services. Early GSM networks could support a dial-up data channel running at 9.6 Kbit/second—the equivalent of a rather slow modem link. This was adequate for exchanging small amounts of data with people on the move but it was totally unsuitable for more demanding applications. The mobile networks addressed this issue in the same way as their fixed network counterparts—they built a data capability in parallel with their existing voice capability. In the case of GSM networks, the new data capability was called General Packet Radio Services—or GPRS, for short. GPRS provided mobile users with an always-on data service that could theoretically run at speeds up to 160 Kbit/second.[22] Enhanced GPRS (EGPRS) subsequently increased this maximum data rate by a factor of three.[23] EGPRS is unofficially referred to as a "2.75G" technology[24] because it is now seen as an evolutionary path to 3rd Generation ("3G") mobile networks rather than simply as a fix for the data weaknesses of GSM ("2G") networks.

The primary objective of 3rd Generation mobile is to provide a single architecture that can carry voice, data and other forms of traffic efficiently. Since 3G was intended to be a quantum step forward rather than just an enhancement to existing networks, it is perhaps not surprising that it was necessary to move to a completely new radio access technology that was incompatible with existing 2G base stations and handsets.[25] However, 3G handsets must be able to operate on 2G networks in areas where no 3G coverage is available. Fortunately, the standards make provision for an evolution path in which the core network can support a mixture of 2G and 3G radio access. Eventually, the need for 2G access will disappear and a single packet switched network will support all mobile voice and data services.

Depending on your point of view, the transition to the LTE (Long Term Evolution) standard will either represent the final stage in the evolution of 3G networks or the start of a 4th Generation of mobile networking technology. As in the case of the transition from 2G to 3G, a new radio interface is going to be needed and it will be completely incompatible with existing mobile handsets.[26] Assuming that the network operators are willing and able to make the necessary investment, LTE is expected to deliver up to 100 Mbit/second on the downlink and 50 Mbit/second on the uplink. These kinds of speeds are capable of delivering high-quality video, so mobile television and video on-demand services could become commonplace, along with applications such as video conferencing and multi-user networked gaming. However, a more significant outcome could be that many customers finally decide to abandon their fixed-line telephone and broadband connection altogether and to use their mobile phone for everything.

This would seem like a substantial victory for the mobile networks in their battle to win customers from their fixed-network competitors. On the other hand, such battles may start to become increasingly irrelevant as technology developments blur the distinction between fixed and mobile networks. Mobile network standards have been incorporated into the core of next-generation fixed networks (such as BT's "21st Century Network"), suggesting that mobile and fixed will simply be different ways of accessing the same network—rather than two completely separate networks. We will return to this discussion in Chapter 13 when we consider the future of telecommunications.

\* \* \*

Before leaving the subject of mobile networks, we should mention that there is one type of mobile phone that can operate in areas that have no cellular network coverage. As its name suggests, the satellite phone connects to a network via a satellite rather than via a cellular base station. Since the satellite is considerably further from the handset than a base station would be, satellite phones tend to require a large retractable aerial and they typically do not work indoors. Call charges are higher than on conventional mobile phones, so usage tends to be restricted to people—such as mariners or oil company employees—who need to operate in remote areas. Furthermore, satellite networks are based on proprietary standards, so they have not benefited from the economies of scale generated by successful standards such as GSM. As a result, satellite phones tend to be larger and heavier than their mobile equivalents.

Some satellite phone networks are based on geostationary satellites,[27] while others use Low Earth Orbit (LEO) satellites. Geostationary satellites are placed in a particular orbit that keeps them in a fixed location in the sky and it is theoretically possible to provide global coverage with just three satellites. However, geostationary satellites have to orbit above the equator, so they appear to be very close to the horizon when viewed from polar regions and could be obscured by relatively minor features in the landscape. Since these satellites are located 35,786 kilometers above the Earth, a radio signal takes roughly a quarter of a second to get there and back, leading to a noticeable delay for the user. The Inmarsat and Thuraya satellite phone networks use geostationary satellites.

Delay is much less of a problem for LEO satellites, because they orbit much closer to the Earth's surface.[28] However, this means that the satellites do not remain in a fixed location in the sky, so a "constellation" of satellites is needed to ensure that a new satellite appears above the horizon before the previous one has disappeared. LEO satellites tend to be cheaper to build and launch than geostationary satellites, but more of them are required to provide uninterrupted coverage. LEO satellites are not restricted to orbiting around the equator, so they can provide better coverage in polar regions. The Iridium and Globalstar networks use LEO satellites.

# 12 When failure is not an option

At 2 a.m. on March 29, 2004, a fire broke out in a tunnel deep below the streets of Manchester. Firefighters wearing breathing apparatus had to climb down a 30-meter shaft and make their way along 50 meters of smoke-filled tunnel to reach the scene of the blaze. Unfortunately, the tunnel linked the BT exchanges at Dial House and Rutherford House, and the fire damaged a large number of critical fiber optic cables. One hundred and thirty thousand telephone lines and Internet connections were lost, causing massive disruption to homes and businesses in the Manchester area. The Greater Manchester Ambulance Service reported that its radio network was affected and some 999 services were also hit. People who tried to use their mobile phones found that the problem even extended to some of the mobile networks. Customers lost service for up to a week, triggering substantial compensation claims. During the outage, it was estimated that the cost to local businesses was running at £4.5m a day.

No network will ever be completely immune to fires—or power failures, or floods, or cable cuts, or equipment faults. Unfortunately, the consequences of these problems are becoming increasingly severe. Dot.com businesses and large telephone call centers are obviously critically dependent upon reliable communications, but it is hard to find any business that would not be seriously affected by the loss of its telecommunications links with the outside world. As a result, network reliability has become a key competitive issue, and networks that fail to restore service quickly after a fault will lose customer loyalty. This chapter considers some of the techniques that can be used by network operators to enhance the reliability of their services.

Let's start by drawing an important distinction between Reliability and Resilience. A Reliable network is one with a low level of faults. Network reliability would normally be achieved by high-quality engineering combined with effective operating procedures. However, even a highly reliable network might not be good enough for some services (e.g. those carrying high-value financial transactions). In these situations, investment is required in network Resilience. A Resilient network is one that continues to deliver services to customers in spite of network faults.

Table 7 illustrates the range of problems that can disrupt the operation of a network.

A. Wheen, *From Dot-Dash to Dot.com: How Modern Telecommunications Evolved from the Telegraph to the Internet*, Springer Praxis Books, DOI 10.1007/978-1-4419-6760-2_13,

**Table 7.** Network problems.

| Problem | Possible causes |
|---------|-----------------|
| Hardware failures and accidental damage | Electronic equipment failures or damage to cables can be caused by vibration, electromagnetic interference, lightning, power supply problems, ice, wind, floods, fires, earthquakes or rodents. Satellites can be damaged by space debris or solar flares |
| Deliberate sabotage | Caused by bombs, computer viruses, cable cuts or arson |
| Software failures | Caused by software bugs or problems that arise during software upgrades |
| Operational errors | Caused by network configuration errors, network planning errors, process problems, errors in records or "finger trouble" (i.e. human error) |

Various techniques used to improve network reliability are described in Appendix O. Perhaps not surprisingly, it is concluded that keeping a network operating reliably is largely a matter of doing things "by the book" and resisting the urge to cut corners. The hardware and software should be properly engineered and properly maintained, while the staff in Network Operations should be properly trained and properly managed.

However, for some customers, even that is not good enough. If you are a financial institution with a digital link carrying transactions worth thousands (or even millions) of pounds per minute, you will care very deeply if that link goes down for just a few seconds. The level of service demanded by some business customers is so high that it can only be achieved by substantial investment in network resilience.

A range of different techniques are used to improve network resilience, but they all exploit the same basic principle. To illustrate how this works, let's assume that a critical service depends upon a network component (such as an item of equipment or a length of cable) that fails on average once every 3 years, and then takes 1 day to fix. (This may sound like the start of a maths exam, but bear with me.) There are roughly 1,000 days in 3 years, so the probability that this component is not working at any given moment is roughly one in 1,000. Now let's assume that we have two identical components of this type and they are connected together in such a way that service will be maintained so long as at least one of them is still working. The probability that service will be lost is no longer one in 1,000, but one in a 1,000,000. For many telecommunications applications, this is considered to be an acceptable level of risk.

Think of it like the spare wheel in a car. Although punctures are not an everyday occurrence, they occur frequently enough to make it worth carrying a spare tire. However, the risk of a second puncture before the first one has been repaired is so small that most people do not consider it necessary to carry a second spare tire. After all, each spare tire costs money to buy and its weight results in additional fuel consumption. However, even very unlikely events do occasionally happen. Someone driving across a desert might well decide to carry a second spare tire because although the probability of two punctures is very small, the consequences could be life-threatening.

In other words, we need to understand what the consequences of failure would be before deciding what level of resilience is appropriate. We can never achieve 100% reliability—perfection is not an option in this imperfect world— but it is possible to make the probability of failure as small as we like by adding more and more back-up components. However, the benefits of network resilience come at a price; additional equipment has to be provided to back up each vulnerable part of the network and this "redundant" equipment is not earning its keep when everything is working normally. It is only worth paying for additional resilience if the consequences of failure would be unacceptably painful.[1]

This point is illustrated by an incident that occurred on Telecom New Zealand's network. In June 2005, rats gnawed through a fiber optic cable on a bridge north of Wellington. This wasn't an immediate problem, because the services were successfully switched to an alternative route. Unfortunately, however, the back-up route was accidentally cut by a mechanical digger before the original cable had been repaired. This affected telephone, Internet, mobile and other services for about 100,000 customers. Financial transactions were disrupted and the New Zealand stock exchange was forced to close for most of the day. After the incident, there was considerable debate about whether it was acceptable for a network to be vulnerable to two simultaneous failures. In view of the consequences, the answer in this particular case would appear to be "no".

\* \* \*

So, how can we make a network more resilient? Surprising as it may seem, it is possible to design equipment that will carry on operating even though some of its internal components have failed. Complex bits of kit such as telephone exchanges are built up from a number of separate sub-systems and each active sub-system can be paired with an identical back-up sub-system that is ready to take over if things go wrong.[2] If the equipment contains a large number of identical sub-systems, then there is no need to provide each of them with its own back-up—a single back-up can be shared by the group and it will replace whichever member of the group fails first. The probability that a second sub-system in the group will fail before the first one has been fixed is very small.

A similar form of redundancy is often used in the case of power supplies. A piece of telecoms equipment might be fitted with five power supply units even though it only needs four. Each power supply unit will only be required to deliver 80% of its maximum capacity while all the units are working properly. However, if one of the units were to fail, then the load would be picked up by the four remaining units operating at full capacity.

Modern networks use a large number of computers in one form or another, so resilience techniques are needed to protect the network from the effects of computer failure. Duplicated processors were introduced into telephone exchanges many years ago to improve resilience[3] and fault tolerance techniques developed by the computer industry can be used to protect against hardware failures in network management and operational support systems. Many server vendors now offer machines that contain redundant processors, disks and power supplies, and server clustering techniques enable other machines in a cluster to take over the tasks of a failed machine. The use of distributed databases, combined with regular back-ups, can protect vital information against the catastrophic loss of a data center.

If redundant hardware can be used to protect against equipment failures, what can be done to protect against damage to cables? There are a number of possible answers to this question, but they all exploit the principles of redundancy. For example, telephony traffic from a local exchange can be spread across multiple trunks taking different routes into the core network, as illustrated in Figure 50.

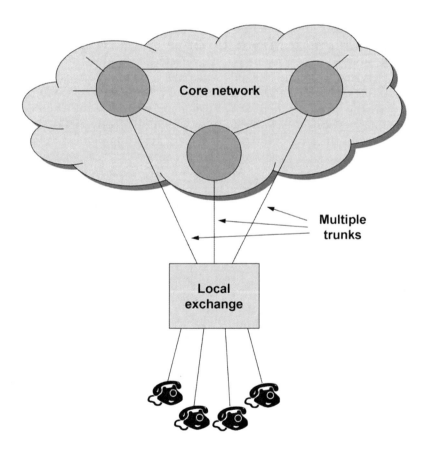

**Figure 50.** Local exchange connections into core network.

If one of the trunks fails, then the calls carried by that trunk will be dropped but calls on the other trunks will be unaffected. As customers re-dial to re-establish dropped calls, they will be routed to one of the remaining operational trunks. Of course, the total capacity linking the local exchange to the rest of the network has been reduced, but customers should not notice any difference if sufficient spare capacity has been provided.

Another simple way in which redundant capacity can be used to protect network traffic is illustrated in Figure 51. This technique is referred to as "dedicated protection". The data going from A to B is simultaneously transmitted on both the upper (primary) path and the lower (back-up) path. The switch at the receiver has been set to use the upper path, so data arriving via the lower path is simply ignored.

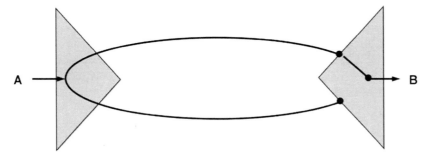

**Figure 51.**   Dedicated protection.

Now let's suppose that the upper path is accidentally cut, as illustrated in Figure 52. The switch at the receiver will notice that data has stopped arriving on the upper path and will switch automatically to the lower path to restore the service. If the two paths are physically separated by a reasonable distance, then it is very unlikely that both cables will be damaged by the same incident. It should be noted that the diagram only illustrates data flowing from A to B—data flowing in the opposite direction is protected by a mirror image arrangement with a switch at A.

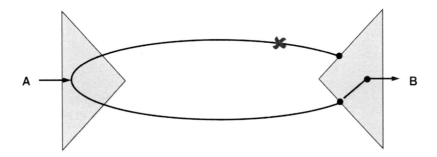

**Figure 52.**   Recovery after link failure.

Dedicated protection techniques such as this can be an effective way of ensuring that the connection between two network nodes is not lost after a cable cut. However, since two separate cables are required, it is obviously an expensive solution. Network restoration provides a slower but cheaper solution for mesh networks such as the one illustrated in Figure 53.

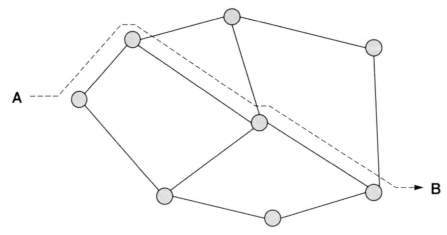

**Figure 53.**   Traffic flowing across a mesh network.

Traffic flowing from A to B can follow a number of different routes through the network. Generally speaking, the route chosen attempts to minimize the number of nodes that have to be passed through while avoiding any links where there is traffic congestion or a history of poor performance. If one of the nodes or links in the mesh were to fail, then traffic would be re-routed around the failure as illustrated in Figure 54.

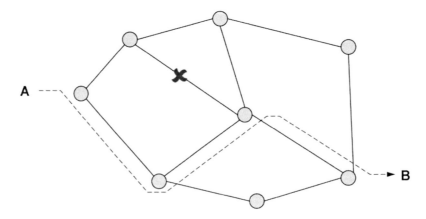

**Figure 54.**   Service restoration after link failure.

Network restoration works by finding alternative routes for traffic affected by the failure. Once the management system has determined how to recover from the situation, reconfiguration instructions are sent out to all affected nodes and the new configuration is activated. The time taken to do this varies considerably between networks, but it may be measured in minutes rather than seconds. This is considerably slower than dedicated protection, which can normally operate in under 50 milliseconds.[4]

At first sight, it may appear that network restoration is giving us something for nothing. Where are the back-up paths that are required by dedicated protection schemes? The answer is that network restoration uses spare capacity wherever it can find it in the network. It is very unusual for network links to operate at full capacity because some headroom is always required to handle unexpected traffic peaks. However, if the spare capacity in part of the network is too small to handle all the re-routed traffic, then it may be necessary to sacrifice some low-priority traffic to make space for more important traffic.[5,6]

The techniques that we have discussed so far were designed to improve the resilience of traditional circuit-based networks. However, packet-based networks can also provide a high level of resilience—indeed, we saw in Chapter 9 how Paul Baran's pioneering work on packet switching was motivated by the need to produce a defense network that would be quite literally bomb-proof. Packet networks actually *expect* packets to be lost or corrupted, so they are well able to handle these problems when they occur.[7]

Packet networks can achieve high resilience by providing plenty of alternative paths across the network and by avoiding any form of centralized intelligence that would be vulnerable to failure. Packet routes across the network are determined by a series of small decisions made by individual routers. Each of the routers is continually chatting to its immediate neighbors, so it will soon become aware if there are any problems in the network. If a router learns that another router has failed or that a link has been cut, then it will change its routing decisions accordingly. Even if whole areas of the network have been damaged, the packets will continue to flow so long as at least one route remains available.[8]

\*   \*   \*

Although a cable cut or an equipment fault is serious enough, networks are also vulnerable to much more devastating problems. A disaster is defined as a catastrophic loss of network capacity and is characterized by a substantial amount of damage. The time required to re-instate the lost network capacity is usually measured in months or even years and so is far longer than even the most tolerant customer could accept. In most cases, a large number of customers are affected by the disaster, leading to significant media interest and severe damage to the reputation of the network operator.

A typical disaster might consist of a building housing large amounts of network equipment being reduced to a smoking hole as the result of an explosion or fire. If the building contains a core network node, then it will usually be possible to re-route traffic around the affected node, although the overall capacity of the network will obviously be reduced. If, however, the building houses a local telephone exchange, then the customers served from that exchange face an extended loss of service. In some cases, the digital circuits used by mobile network base stations will also pass through the same building, so the unfortunate customers may lose not only their fixed-network services, but also their mobile-network services as well. This has obvious implications in emergency situations.

To address this risk, critical business customers can be served from more than one local exchange if the value of their traffic is sufficient to justify the cost. The threat to the mobile network can be reduced by routing the traffic from each base station through a local exchange that serves an area outside the cell coverage area or by using a completely separate network. Some network operators also have a local exchange mounted on the back of a truck so that they can restore basic telephony services relatively quickly after a disaster. Of course, all of these solutions entail significant cost, but the losses resulting from a disaster can be devastating. Customers need to plan for a disaster *before* it happens!

# 13   What comes next?

The telecoms market has changed dramatically since the early 1980s. In those days, fixed telephone services were normally only available from a state-owned monopoly, while mobile services were hardly available at all. Today, there is competition in both fixed and mobile telephony and prices have fallen dramatically. The circuit-switching concept, which has been at the heart of network architectures since the days of Alexander Graham Bell, is giving way to packet switching. Data has overtaken voice as the dominant form of traffic on most networks and many of the services now available on the Internet would have been unimaginable just a few years ago. Telecoms is no longer restricted to communication between human beings—increasingly, it's about humans talking to machines or machines talking to other machines. Even household appliances such as refrigerators are starting to need their own IP address.

So what does the future hold? Will the rapid change that we have seen in recent years be maintained or are we moving from a period of revolution into one of more gradual evolution? Despite the obvious danger of being proved wrong, let's attempt to predict some themes for the next few years.

## 13.1 Next generation networks

There is no agreed definition for a "next generation network", but some of the key characteristics are:

- services carried digitally end to end;
- designed to carry all forms of traffic (voice, data, video, etc.);
- based upon packets rather than circuits;
- using optical technologies to deliver very high bandwidth;
- serving both fixed and mobile users;
- providing much clearer separation between the network and the services that run over it.

Let's examine each of these characteristics in a bit more detail.

The transition from analog to digital began in the early 1980s. Although many core networks are now digital, most domestic telephone lines remain stubbornly analog. It has been estimated that it could take until 2030 before the conversion to digital is essentially complete.

In the past, large operators often had to support six or more separate networks that delivered different groups of services. These networks could, for example,

A. Wheen, *From Dot-Dash to Dot.com: How Modern Telecommunications Evolved from the*
*Telegraph to the Internet,* Springer Praxis Books, DOI 10.1007/978-1-4419-6760-2_14,
© Springer Science+Business Media, LLC 2011

include the voice network, an Internet/IP network, a frame relay network, an ATM network, a low-speed leased-line network and a high-speed leased-line network. Each of these networks would probably have separate systems and processes to keep the network running and each would need a separate workforce. Today, carriers around the world are trying to merge their various networks into a single "converged" network that can carry all services. A converged network can be used more efficiently because capacity can be re-assigned dynamically from one type of traffic to another to meet the needs of the moment. Furthermore, having one set of staff to support a single network is cheaper than maintaining a number of separate teams.

Developments in digital technology and the growth of the Internet have caused the volume of data traffic to rise dramatically. Data is once again the dominant form of traffic on most networks[1] and the gap between data and voice is growing wider all the time. Next generation networks need to be data networks that can carry voice—rather than the other way around—and this is driving revolutionary changes in network architecture. Circuit switching is in decline and will eventually disappear. Networks of the future will be packet-based.

In parallel with the transition from circuits to packets, we are also seeing a transition from electrical to optical networks. Most core networks are now based upon optical fiber, and fiber is becoming increasingly common in the access network. The huge bandwidth capacity of optical fibers has allowed networks to keep pace with growth in data traffic, but electronic equipment is still found at the ends of an optical fiber so a signal will still experience frequent conversions from optical to electrical and back again as it crosses a network. An all-optical network would eliminate the need for many of these conversions, so the search is now on for technologies that will allow packet switching to take place entirely in the optical domain.

Mobile networks did not appear until about a century after the fixed telephone network, so it is hardly surprising that the two networks developed separately—and often in direct competition. However, the distinction between fixed and mobile networks is now starting to disappear. Next generation networks will be accessed using devices ranging from smart phones to games consoles and from household appliances to digital cameras. Who cares whether a device is fixed or mobile? So long as it meets the needs of the customer, that's all that matters.

Next generation networks will carry a full range of services—including some that have not yet been invented. This has required a much clearer distinction between the basic, "bit-shifting" functionality of the network and the range of services and applications that are built on top of it. Development of new network applications will become simpler and this will stimulate the development of applications by organizations outside the telecoms industry in much the same way as the launch of Apple's iPhone triggered dramatic growth in third-party applications for mobile phones.

Many of these new applications will gobble up large amounts of bandwidth, so next generation networks will not achieve their full potential if the access network continues to represent a bandwidth bottleneck. Fiber To The Home will eventually become essential for fixed network access, while 4th Generation technologies will be needed for mobile access.

## 13.2 Green networks

Networks consume substantial amounts of electricity. As the world becomes increasingly concerned about the effect of carbon emissions on the climate, we are starting to see a new focus on reducing the power consumption of network equipment. Whilst it can be argued that networking reduces the need for many carbon-intensive activities (e.g. videoconferencing reduces the need to travel), this is not an excuse for inaction. Solar and wind-powered base stations are already providing mobile coverage in remote parts of the world and solar-powered mobile phones are starting to appear.

However, the networking industry has not kept pace with the power reductions that have been achieved in the IT sector. The widespread adoption of virtualization technologies has meant that software applications that previously required their own dedicated server can now run quite happily in a "virtual machine" and a large number of these virtual machines can share a single physical server. As a result, existing applications can be consolidated onto fewer physical servers and the remaining servers can be switched off. If usage fluctuates during the course of a day, then additional servers can be switched off at times at which they are not needed.

It is tempting to think that virtualization techniques could be used to reduce the capacity of networks at times at which demand is low, but there is not much evidence that this is happening. Whilst the power consumption of some network equipment fluctuates in response to the level of traffic that it is carrying, we do not often see parts of a network being powered-down at times of low demand. This is an area that appears to offer scope for innovation.

## 13.3 Mobility

In 2002, the number of mobile subscribers worldwide exceeded the number of fixed-line subscribers for the first time. The development of mobile voice and data services has been one of the great success stories of recent years, with the mobile networks pioneering a range of innovative services such as text, picture messaging, personalized ringtones and pre-payment with electronic top-up. Of course, some of their greatest successes have been happy accidents—text messaging was originally developed to transmit engineering information, but became successful when teenagers discovered that it was much cheaper than voice calls. Conversely, some carefully planned marketing initiatives have been flops. However, we can reasonably assume that the rapid innovation seen in recent years would not have occurred if the telecoms market was still controlled by bureaucratic state-owned monopolies.

Many countries now have an average of at least one mobile phone for every member of the population, so mobile network operators have had to grow by stealing customers from the fixed networks or—to put it more politely—through "fixed-mobile substitution". As an example, fixed network calls are normally cheaper than mobile calls, so mobile customers have been offered fixed-network tariffs for mobile calls made from home.[2] Fixed–mobile substitution has been relatively successful (particularly among young people and those with no permanent address) and fixed telephones have now disappeared from some homes.

However, just at the point at which the copper telephone line might have disappeared into history, the arrival of broadband services gave it a new lease of life. Furthermore, the need to improve in-home mobile coverage led to the development of femtocells and these require a broadband connection. It is not that easy to do without a fixed-network connection! We can therefore expect fixed-mobile substitution to give way to "fixed-mobile convergence". While fixed and mobile services continue to be delivered by separate networks, customers are required to have separate telephones, separate phone numbers, separate messaging systems, separate electronic address books and separate bills. Fixed-mobile convergence aims to give users the best of both worlds—the convenience of the mobile network with the high capacity and low cost of the fixed network.

In recent years, call charges on fixed networks have been driven down remorselessly by the effects of competition and by the falling cost of bandwidth. In spite of this, mobile calls continue to command premium prices and international roaming is even more expensive. Mobile operators have been able to maintain their voice prices by restricting the user's ability to use VoIP, but these restrictions are now starting to break down and we can expect to see the cost of mobile calls falling as they did on the fixed network. The mobile network operators will need to change their business models if they are to avoid becoming nothing more than providers of low-cost mobile data connections that carry other people's money-making services.

A further threat to the mobile operators could come from metropolitan wireless networks based on technologies such as WiMAX or WiFi. These originally appeared as "hotspots" in coffee shops and airports, but coverage has spread across significant areas of large cities and is now available on some trains. There is a risk that these networks could help themselves to voice and data traffic that mobile operators would previously have regarded as their own. Furthermore, these wireless networks can focus their coverage on high-value locations, while the mobile networks have regulatory obligations to provide much wider network coverage. In the jargon of the industry, wireless networks can be used for "cream skimming". The mobile operators have done extremely well from the world's addiction to mobile communications, but the party may be coming to an end.

## 13.4 Terminals and handsets

Once upon a time, the only device that you could connect to a telephone network was ... a telephone. Next generation networks will have to work with a bewildering variety of gadgets and many of these will turn out to be mobile phones in one form or another.

The mobile phone evolved from early devices that were far too big and heavy to carry around, but they could be installed in a car. The extraordinary reductions in size and weight that occurred as the technology matured have enabled the carphone to evolve into a personal mobile device that is truly portable. Furthermore, it is no longer simply a telephone—a mobile can also be a camera, radio, MP3 player, navigation aid, computer, Internet access device and a terminal for cashless payments. For some users (including those in the developing world, where computers are not widely available), the mobile handset has also become the primary computing environment.[3]

On the other hand, for those mobile customers who confine themselves to phone calls and text messaging, the mobile phone has become an electronic version of the Swiss army knife—highly featured, but larger and heavier than it really needs to be. For these people, the concept of a mobile phone that is so thin and flexible that it can be worn as a wrist watch when not in use could represent an attractive alternative.

So, how will the mobile phone evolve over the next few years? It is a pretty safe bet that there will be steady improvements in display technology, storage capacity, battery life and user interfaces, but predicting which new applications will be successful is a little harder. Let's focus on just three of them: augmented reality, mobile payments and mobile TV.

Most mobile phones already include a camera. Many also know where they are and in which direction they are pointing. These capabilities can be combined with pattern recognition to enable the phone to recognize features in its environment. For example, the phone could be used to identify a particular statue in a city centre and to provide information about the sculptor. In many cases, the information would be superimposed on the image to provide an augmented form of reality. This technique is already seen on sports broadcasts, where the commentator can explain how a goal was scored by drawing the path of the ball on top of the video image.

Mobile network operators have established methods of charging their customers for phone calls, so the same billing mechanisms could also be used to pay for other goods and services. In one early experiment, customers were able to buy bars of chocolate from a vending machine by calling a particular phone number from their mobile. More recently, payment systems based on text messaging have appeared. However, the popularity of online shopping has stimulated the growth of web-based payment systems for mobile phones. As in the case of Internet shopping, mobile web payments can be used to purchase downloads (such as music, ringtones and games), services (such as travel tickets and car parking) and physical goods (such as books).

Mobile payments have significant applications in developing countries where few people have access to a bank account or a credit card. They can also be used instead of cash for small transactions, including those "micropayments" that are too small to be handled cost-effectively by credit or debit cards. However, buying a loaf of bread by accessing a website on a mobile phone is not very convenient. Proximity card ("swipe card") technology can already be used to pay for canteen meals and for travel on the London underground and this technology has now been extended for use on mobile phones. Near Field Communication (NFC) provides a much faster and more convenient way of making mobile payments at a point of sale. For small payments, there is no need to type in anything—just swipe your mobile phone and it is done. (For larger payments, a PIN might be required to confirm the sale.) NFC also has a range of other potential uses. It could, for example, be used to leave contact details without the need to carry business cards or to provide access to restricted areas of a building.

Unfortunately, there is a snag. NFC will not be widely used until a large number of retailers have compatible terminals in their shops and this is unlikely to happen until a large number of customers have NFC-compatible phones. On the other hand, users are unlikely to pay extra for NFC if they cannot use it immediately. Fortunately, there is a way out of this vicious circle. Some of the advanced capabilities provided on mobile phones are not there to meet existing consumer demand—they are there to see whether new consumer demand can be created. In the case of Bluetooth, demand did not take off until a significant number of existing handsets supported the technology and it appears that the same approach will be used to promote NFC-based mobile applications. If this happens, a whole new range of applications could open up for the mobile phone. Not only will it be a camera, a computer and an MP3 player—it could also be a wallet, an ID card and a set of keys!

It could even become a television set. Recent experiments with mobile TV have produced mixed results, but it seems likely that the mobile TV concept has a viable future if the right business model can be found. For example, will users want to watch streamed video on their mobiles or will they prefer podcasts?[4] How much will they be prepared to pay? Should mobile TV offer standard television shows that have been modified for the (very) small screen or do TV programs need to be created specifically for mobile TV?

In addition to the business issues, there are also a number of technical challenges. A simple method of transmitting broadcast television to a mobile phone would be to stream the signal over a 3G mobile network. However, this can be a rather wasteful use of network bandwidth because—at least in principle—it can result in a separate broadcast stream being transmitted across the network to each separate viewer. In contrast to this, your television aerial receives the same signals as your neighbors' television aerials, so there is no need to transmit each TV channel more than once. This has led to a number of experiments in which conventional television transmitters are used to broadcast to mobile phones equipped with suitable decoders. Unfortunately, there are competing technical standards[5]—each with strengths and weaknesses—and the lack of international agreement is reducing the potential for volume-related cost reductions.

These issues will take time to resolve, but the success of personal entertainment devices such as Apple's iPod suggests that mobile TV may eventually find a role. Young people will almost certainly be early adopters of the technology and 45% of the world's population will be under the age of 24 by 2015. In many parts of the world, these people will have grown up surrounded by digital technology and will probably have owned a mobile phone from an early age. For these people, the mobile phone will eventually provide a remote control for almost every aspect of their lives.

## 13.5 Service providers

The Internet is essentially "dumb", with all the clever stuff being implemented in servers sitting around the edges of the network. This architecture means that services available on the network do not have to be controlled by the network operator—anyone can do it! Skype, Google, Facebook and the BBC iPlayer are all examples of popular Internet-based services that are controlled by independent service providers. This, of course, is potentially bad news for the network operators. Ownership of the network no longer gives them complete control over the services that the network delivers and some network operators have seen lucrative parts of their businesses skimmed off by service providers. In many cases, the network operator is no longer visible to the end user because the customer relationship is controlled by an independent service provider. As a result, network operators have been relegated to low-value activities such as moving bits around the network. Furthermore, the service providers are pushing up the demand for bandwidth, which increases the network operator's costs without necessarily increasing their revenues.

Naturally, the network operators are trying to resist these changes and the development of next generation networks provides them with an opportunity to introduce new value-added services of their own. Many of them would like to move away from the concept of a "dumb" network towards a network with a much higher level of embedded intelligence. The network would provide a range of capabilities that could be used like building blocks to create new services. For example, if the service requires knowledge of the user's physical location, then this would be supplied by intelligence embedded within the network. BT's "21st Century Network" is an example of a network operator trying to resist the dumbing-down of its network.

The relationship between service providers and network operators can be mutually beneficial if the service providers can sell into market segments (such as teenagers or ethnic minorities) that traditional network operators find hard to reach. This has led to the emergence of Mobile Virtual Network Operators (MVNOs) such as Virgin Mobile and Tesco Mobile. These service providers have strong consumer brands but they do not have a network of their own; instead, they sell services on someone else's network. This requires the cooperation of the network operator, but the network operator benefits from the additional traffic that it brings to their network. Virtual network operators can offer both fixed and mobile services and some have focused on the needs of business customers. The virtual network operator concept has generally worked well and we can expect it to be more widely adopted in the future.

## 13.6 Quality of Service

The transition from circuit-based to packet-based networks has led to a number of Quality of Service (QoS) issues and next generation networks provide an opportunity to address these issues. In a circuit-based network, bits are transmitted at a fixed rate (such as 64 Kbit/second for voice services). All the bits using a particular circuit follow the same path across the network and they are unaffected by network congestion because the circuit bandwidth is guaranteed.[6] The low and predictable transmission delays offered by circuit-based networks are ideal for delay-sensitive applications such as voice.

Of course, packet-based networks can carry voice services, but the quality is sometimes less than wonderful—as users of Internet-based voice services will confirm. Various techniques have been devised to address this problem. On some networks, delay-sensitive packets are sent straight to the front of every queue, while less critical packets are made to wait their turn. Other networks take a more "belt-and-braces" approach by providing so much bandwidth that congestion simply is not a problem. However, problems can still occur on interconnected networks (such as the Internet) because the quality of an end-to-end link is only as good as its weakest network. Next generation networks will be able to guarantee end-to-end performance in return for a premium price, thereby providing an opportunity for network operators to increase the value of their services. We can expect to see alliances between these networks to provide high-quality services to corporate customers on a global basis. Some people believe that this could lead to the emergence of a two-tier Internet.

Although packet prioritization techniques are primarily intended to improve network performance for delay-sensitive traffic, they could also be used to degrade performance for traffic that the network operator wishes to discourage. They could, for example, be used to ensure that the network operator's own voice over IP services perform better than competitors' offerings or to discourage traffic (such as video and peer-to-peer) that can increase an ISP's bandwidth costs without generating additional revenue. This has led to the "net neutrality" debate. Is it right, for example, for an ISP to enhance the quality of access to a particular website that pays them a fee? On the other hand, should lucrative websites such as Google be able to hitch a free ride on expensive network infrastructures? Are all bits created equal or are some more equal than others? This may seem like an arcane debate, but it could have significant implications for the future of the Internet. Strong views have been expressed on both sides and the final answer is not yet clear.

## 13.7 Internet applications

Predicting future developments is particularly difficult in the case of the Internet. Not only do new Internet applications appear with bewildering speed, but many of the most successful applications have popped up unexpectedly. This has been happening since the earliest days of the ARPANET[7] and there is no reason to expect that it will change any time soon. Although we cannot predict the "next big thing", we can identify some general trends that are likely to continue.

The arrival of the World Wide Web has disrupted traditional business models in many different industries. The costs of doing business online are generally low and even the smallest of companies can establish a global presence. This has made it possible to address niche opportunities that were beyond the reach of conventional business models. For example, Amazon sells many specialist books for which there is so little demand that no high-street bookshop could afford to keep them in stock. Although each individual book brings in very little revenue, the cumulative effect is a substantial business.[8] We can be pretty confident that the Internet will continue to disrupt conventional ways of doing business and will enable companies to expand beyond the confines of their traditional markets.

We can also expect no let-up in the rate of innovation. Standards-based software interfaces are making it increasingly simple to experiment with web-based concepts by linking together existing applications in new and innovative ways. To illustrate just how quickly such "mashups" can be created, a small team of programmers recently developed an application to convert the soundtrack of a television soap opera from English to German.[9] The conversion process had four main steps:

1. Extract the subtitles from the digital television signal.
2. Convert the subtitle text from English to German using Google Translate.
3. Convert the German text to speech using a speech synthesizer.
4. Synchronize the German speech with the video signal.

This new application took less than 1 day to develop. Software already existed to perform the key steps in the process but the clever part was to link the software together using standard programming interfaces. The robotic speech produced by the speech synthesizer meant that the German soundtrack carried considerably less emotion than the original English, but prototypes created in this way can be refined using more traditional programming techniques if the additional investment proves to be justified.

In the past, innovation on the Internet was seriously constrained by lack of bandwidth. However, advances in optical networking have eased these constraints and this has prompted comparisons with the way in which advances in semiconductors enabled developments in the personal computer industry. Just as more powerful personal computers led to increasingly sophisticated applications with a wealth of new features, so more network bandwidth will encourage the use of bandwidth-intensive forms of content. If the analogy with PC software is a reliable guide, some "enhancements" to existing services will be ignored or actively resisted by those users who prefer to keep things simple. In spite of this, we can be reasonably sure that new network bandwidth will eventually be filled up one way or another, even if we don't yet quite know how.

## 13.8 Cloud computing

One development that could drive the growth of network traffic over the next few years would be the migration of application software (word processors, spreadsheets, etc.) from personal computers to Internet-based servers. To understand why this might happen, imagine the problems currently faced by an administrator who is responsible for thousands of laptop computers belonging to an organization. Each computer has a range of different applications installed and each application has to be properly licensed. Furthermore, software upgrades and bug fixes are required periodically and there is a constant need to ensure that anti-virus and security software is kept up-to-date. Managing these issues for large numbers of laptops that may be widely dispersed around the country (or even the world) can be a nightmare!

However, a high proportion of homes and offices have broadband Internet access and wireless technologies such as WiFi, WiMax and 3G are making it increasingly unusual for computers to be disconnected from the Internet. This makes it possible for common PC applications to be operated and managed centrally on an Internet server rather than being installed on each separate computer. We already use Internet-based applications such as search engines, so why should word processors or spreadsheets be any different?

We mentioned in Chapter 10 that it is becoming increasingly difficult to draw a clear distinction between a network and the computers that are connected to it. This has now given rise to the concept of "cloud computing". The name is a reference to the traditional diagrammatic representation of the Internet as a fluffy cloud, but cloud computing is based on the assumption that the cloud contains not only a network, but also large amounts of processing power, storage and applications software. Users connected to the cloud have access to a full range of services but they do not need to know where those services are coming from or how they are implemented. If this idea is taken to its logical conclusion, a substantial amount of software (including most of the operating system) would migrate from personal computers into the cloud. The personal computer would then become little more than a dumb terminal providing access to facilities offered by the cloud.

Common office applications are already available online and many of them aim to recreate the familiar desktop environment. Gone are the days when you had to upgrade your relatively new laptop because it was no longer powerful enough to run the latest software—any upgrades required occur inside the cloud. In fact, there is no need to carry a laptop around at all because cloud-based applications and data can be accessed from any computer with an Internet connection and even a mobile phone is capable of controlling power-hungry software. Collaboration on a document or presentation becomes much simpler because every member of the team has access to the latest version. The idea of "software as a service" is not new, but the increasing ubiquity of high-bandwidth Internet access may mean that its time has finally come.

On the other hand, the migration of software and its associated data from a physical computer to somewhere in the cloud raises some pretty obvious concerns about security. We have all heard about cases in which laptop computers containing highly sensitive information have been lost or stolen, but maintaining security on an Internet server whose physical location may be completely unknown presents a whole new set of challenges. There is no suggestion that the technologies to address these challenges do not exist, but even the best technology can be compromised by process failings or human error. It remains to be seen whether these concerns can be addressed sufficiently to allow a wholesale migration to cloud computing.

## 13.9 Content

Within the telecoms and broadcasting industries, the term "content" typically refers to material that is watched, listened to or read. A television program, an MP3 download, a web page, a movie or a newspaper article are all forms of content. When they are not being used to provide communication between people or machines, telecoms services are often being used to deliver content in forms such as podcasts, premium-rate phone lines and IPTV. In many cases, the value of the content is considerably higher than the value of the telecoms service used to deliver it.

The traditional methods of television broadcasting are all based on the idea of a single, professionally produced piece of content (a television program) being delivered simultaneously to a large number of viewers. In the United Kingdom, analog terrestrial broadcasting was restricted to five channels (BBC1, BBC2, ITV, Channel 4 and Channel 5) by the limited radio spectrum that was available, so each channel had to appeal to a wide audience. The transition from analog to digital terrestrial broadcasting has enabled this spectrum to be used far more efficiently and there are currently more than 40 channels on the United Kingdom's Freeview platform.[10] Satellite television, which operates at higher radio frequencies, can support hundreds of separate television channels. This is changing the economics of the industry. Most television channels are funded primarily by advertising revenues and advertising slots at peak viewing times were extremely expensive in the days when only five channels were available. As the number of channels has increased, the audiences for each individual channel have declined and this has had a knock-on effect on advertising revenues. On the other hand, the audiences for each channel may have become better segmented. For example, viewers of a history channel presumably have at least one thing in common—an interest in history—and so this channel might be a good place to advertise books, movies or holidays with a historical theme.

The Internet has taken this segmentation process a step further. Technologies such as podcasting mean that it is now viable to broadcast to arbitrarily small groups of people with highly specific interests. You could have a podcast serving a village book club, but it is possible to imagine even more extreme market niches. Why not, for example, have a podcast for Polish-speaking people who share an interest in collecting antique barbed wire? It doesn't matter if your target group is scattered around the world—so long as they all have access to an Internet connection.

These developments are taking us well away from the slick, professionally produced television and radio programs that we have been used to. Such programs require large audiences to justify their high production costs and these audiences are increasingly hard to find. Of necessity, a high proportion of the content available on the Internet is user-generated, but this has led to an explosion of creativity that can be seen on sites such as YouTube. Of course, some of this material is of questionable quality (and taste), but the Internet has created exciting new channels through which aspiring artists and musicians can showcase their talents. Rock bands can make their debut single available as a free download or can distribute it through peer-to-peer file-sharing networks. Live performances can be staged in Second Life and viewed around the world. Websites, video clips and downloadable ringtones can all be used to promote the band to a wider audience. This explosion in user-generated content can also be seen in the social networking activities that are collectively referred to as "Web 2.0". This term includes sites such as Facebook, Twitter, Wikipedia, YouTube and eBay that would be completely meaningless without user-generated content.

It may be hard to believe but teenagers already spend more time surfing the Internet than they do watching TV. For many, the Internet has become the primary source of information and entertainment, and the extraordinary range of content available is making traditional broadcasting look rather restrictive. Why should a viewer be limited to a few hundred channels when the choice on the Internet is almost unlimited? And why should people have to watch a television program at a time that has been chosen for them by some anonymous television scheduler? Internet content is available whenever they want it.

The Video on Demand services available on cable TV allow viewers to break free from the constraints of "linear" television and create their own personalized television schedules. These schedules will probably include some broadcast content that has simply been time-shifted to a more convenient moment, but it could also contain content that has been downloaded from the Internet. "Metadata" embedded within the content would describe it in terms of subject, genre, leading actors and other criteria, so personal video recorders (PVRs) could detect a viewer's preferences and then download material that is likely to be of interest.

In some cases, the content owner will wish to charge viewers for use of the content. That's not a problem. If your PVR decides that a particular movie or television program is likely to appeal to you, it could speculatively download the content to its hard disk without actually bothering to ask you. Then, when the download is complete, it would offer you the content through the on-screen electronic program guide. Although no money has yet changed hands, the content is encrypted so you cannot actually do anything with it. If you decide to watch it, then an electronic payment is made and the encryption key is downloaded to your machine. This mechanism, known as Digital Rights Management, can be used to distribute content through uncontrolled channels such as peer-to-peer file-sharing networks. The content is safely hidden inside its encryption wrapper until someone decides to pay for it. Even when content has been paid for, the Digital Rights Management may impose restrictions on how long the content is available to you or how many times it can be watched.

Taking things a stage further, why should the content only be viewable on a television set? Why can't it be transferred to a laptop computer or a mobile phone or a portable games console? Transferring content between devices is currently impeded by both technical and commercial considerations, but there is no reason to believe that these issues could not be overcome if the market demand became strong enough.

### 13.10 Television

As we have seen, the switch from analog to digital broadcasting has allowed a substantial increase in the number of television channels available. Even so, the tens of television channels available on terrestrial TV cannot match the hundreds of channels available on cable TV and satellite. As the demand for high-definition TV increases and three-dimensional television starts to appear, terrestrial broadcasting is likely to be squeezed by lack of bandwidth.

However, its competitors also have some problems. Cable TV struggles with the very high cost of establishing and maintaining a cable connection to every subscriber, while satellite TV lacks a built-in return path to support full interactivity.[11,12] A more serious weakness with satellite TV is its inability to support Video on Demand (VoD) properly. As discussed in the previous section, viewers are increasingly demanding the ability to create their own TV schedules and it is clearly uneconomic to dedicate a separate satellite transponder to each Video on Demand session. Installing a hard disk in the set-top box allows content to be downloaded to the box for later viewing, but the viewer has to wait while the download proceeds; they do not get the instant gratification of selecting what they want to watch and then pressing the Play button.

IP Television (IPTV) is the new kid on the block. Although the idea of delivering television services over a broadband connection is not new, it took improvements in video coding (to reduce the bandwidth required by the content) and in ADSL technology (to increase the bandwidth of the connection) before the concept really took off.[13] IPTV has proved highly attractive to large network operators because it enables them to derive new revenue streams from their existing copper-based access networks. Like Cable TV, IPTV can deliver true Video on Demand and it has a built-in return path to support full interactivity. Furthermore, it is possible to deliver television, telephony and broadband (the so-called "triple play" services) simultaneously over a copper telephone line. However, delivering multiple high-definition TV services could push copper close to its bandwidth limits and this, finally, could be the reason for replacing copper cable with optical fiber in the access network. It will take a while, but we can be confident that Fiber To The Home (FTTH) will happen eventually.

## 13.11 Payments

The charge for a telephone call has traditionally been determined by:

- call duration (in minutes or seconds);
- time of day (peak/evening/weekend);
- distance (local/national/international);
- type of call (freephone/premium-rate, etc.).

This form of charging was perfectly rational when network bandwidth was a scarce resource that could not be wasted. However, it makes much less sense in a world in which network bandwidth is plentiful and voice represents an increasingly insignificant proportion of total network traffic. As a result, it is becoming common to find tariffs that offer unlimited calls (with some exclusions!) in return for a fixed monthly payment. This means that bills are simpler to calculate and simpler to understand—so disputes are less likely and everyone is happy.

In some ways, the voice networks are simply copying the charging mechanisms used on the Internet. ISPs typically offer a fixed monthly charge for Internet access, with lower charges available in exchange for a limit on the amount of data that can be downloaded each month. Payments for "peering" arrangements between large network operators in the Internet core are often even simpler—traffic flows in both directions, but no money changes hands.

In the future, we can expect to see more innovative charging mechanisms. Whilst service provider revenues will have to come from somewhere, they won't necessarily come from the end user. Skype, for example, has demonstrated that it is possible to offer a free telephony service over the Internet—they make their money from selling enhanced features such as voicemail and call forwarding and by charging for calls that originate or terminate on the telephone network. Of course, the user still has to pay for Internet access, but even that can sometimes be free—or apparently free. Freeserve was a UK Internet Service Provider that acquired more than two million customers between 1998 and 2000 by offering "free" dial-up Internet access (the cost of the Internet access was effectively covered by revenues from the telephone calls,[14] supplemented by advertising revenues). More recently, "free" broadband services have been offered by Carphone Warehouse (the cost of Internet access was recovered from the bundled voice service that the customer was required to buy as part of the deal).

Many Internet applications are genuinely free to the user. Free email and web search facilities are available from providers such as Google, Microsoft and Yahoo! and vast amounts of information are freely available on websites. In some cases, a basic service is offered free of charge in the hope that some users will eventually decide to pay for the fully featured version (e.g. free anti-virus software). In other cases, advertising provides some or all of the revenues (e.g. Facebook, YouTube). In fact, the Internet has proved to be a very effective place to advertise because the advertising can be specifically targeted to address the needs or interests of a particular user. For years, advertisers have complained that half the money they spend on advertising is wasted—but they don't know which half![15] The extraordinary revenues generated by the Google search engine suggest that Internet advertising has a far higher success rate than traditional methods such as television commercials or billboards.

Network operators are starting to learn the lessons of the Internet. A mobile phone is typically used by just one person, so by keeping track of their likes and dislikes (and combining this with knowledge of their current location and the time of day), the network could provide highly targeted advertising. This could be in the form of a short announcement at the start of each phone call or a short header inserted into SMS text messages. Either way, the user receives a cheaper service.

## 13.12 Security

As networks become more sophisticated, the opportunities to misuse them increase. AT&T discovered this as early as the 1960s, when the introduction of a new signaling system led to an outbreak of phone phreaking.[16] However, the large number of software-based features in modern networks means that the opportunities to misuse them have greatly increased. Even mobile phones are now vulnerable to viruses distributed by Bluetooth and media messaging. Network security has become a pressing problem, and one that is hugely expensive to address.

Computer viruses first appeared in the early 1980s and were propagated from one computer to the next on infected floppy disks. However, the arrival of email and the Internet created new and much more efficient ways of distributing viruses, and led to the development of whole new types of "malware". For example, worms are malicious programs that can copy themselves from one computer to another without the need for user intervention. Trojans hide on an infected computer watching for passwords or they simply hang around until they receive an external command to do something nasty. Perhaps most worrying of all is the Denial of Service attack, which simply bombards its target with messages until it collapses under the strain.

Early forms of malware were mostly created without any malicious intent. Programmers were motivated by a desire to explore the boundaries of what was possible or the need to prove their intellectual superiority. However, as the Internet became an increasingly critical part of business life, its possibilities began to attract the attention of organized crime. Online retailers and other companies whose revenues depend upon their websites can be brought to their knees by a Denial of Service attack and so make tempting targets for extortion.

The messages used in a Denial of Service attack are typically sent with random source addresses, so they cannot easily be filtered out. The Distributed Denial of Service (DDoS) attack is even more difficult to deal with, because it uses a large number of infected "zombie" computers to launch the attack simultaneously. Once a network of zombies has been created, attacks can be turned on or off by simply issuing a command from the controller. The owners of the zombie computers are almost certainly unaware that their computers are being used to launch attacks. The victim, on the other hand, will be very much aware that he is under attack, but will be hard pressed to do much about it.

In 2007, a Soviet war memorial was removed from a central square in Talinn, the capital of Estonia. This led to strong protests by ethnic Russians and triggered a DDoS attack against important Estonian websites that lasted from April 27 to May 18. Although difficult to prove, it appears that this attack may have been launched from Russia, so it could possibly represent the first significant cyber attack by one sovereign state upon another. Estonia makes extensive use of the Internet and the attack brought down considerable proportions of its government, banking and healthcare systems. Cyber attacks will certainly be a feature of future warfare, but they will be launched with a ferocity that will make this episode look like a gentle warning.

Since the attack on the World Trade Center on September 11, 2001, there has been considerable speculation about the possibility of cyber terrorism. In March 2000, a disgruntled Australian employee used the Internet to release one million liters of raw sewage into the rivers and coastal waters of Queensland.[17] Could terrorists use the Internet to gain control of a nuclear power station, an air traffic control system or a military installation? The possibilities for causing mayhem in this way are limited only by the imagination, and the fact that it hasn't happened yet does not prove that it can't be done. Hackers have demonstrated convincingly that it is possible to penetrate highly sensitive government and military computers using nothing more sophisticated than a laptop computer and a connection to the Internet.[18] Is it only a matter of time before the West is held to ransom by terrorists operating from somewhere beyond the reach of legal or military sanctions?

Experts dismiss these suggestions by pointing out that critical systems are "air gapped" from external networks (i.e. there is no physical connection at all). But mistakes are still possible. During testing of the Boeing 767 Dreamliner, it was found that no air gap existed between the aircraft's control systems and the network used to provide passengers with in-flight Internet access. If this vulnerability hadn't been found during a routine FAA inspection, it might have enabled a passenger with advanced hacking skills to have a go at flying the plane.

Another threat to the integrity of the Internet comes from spam. Named after the famous Monty Python sketch ("spam, spam, spam ... lovely spam"), spam is email that is sent out to advertise everything from Viagra to online gambling. It doesn't matter that nearly all the emails are deleted as soon as they arrive; spam can be sent out in such vast quantities—and so cheaply—that a very low response rate is still enough to make it worthwhile. Unfortunately, the Internet is now creaking under the strain and email services without spam filtering are becoming unusable.

"Two years from now, spam will be solved," declared Bill Gates to the World Economic Forum in 2004. Since then, spam has grown from an occasional irritant to a major problem (in October 2007, it was estimated that spam accounted for up to 95% of all emails). Most spam is sent from networks of zombie computers and each email in a batch can have slightly different wording so it is not easy to filter it out automatically. As if that wasn't bad enough, spammers are now hiding their messages in randomly generated images; not only are these much harder to detect, but they fill up considerably more bandwidth than conventional text emails. If this menace cannot be checked, then spam could eventually bring about the end of email as a viable form of communication.

## 13.13 Health concerns

The health risks associated with radioactivity have been recognized for many years. Alpha, beta and gamma emissions from radioactive sources can ionize atoms or molecules that they encounter by detaching electrons from them. People exposed to high levels of ionizing radiation can develop a number of health problems, including radiation sickness, sterility, cancer and genetic damage. Since gamma rays are a form of electromagnetic radiation, it is not surprising that concerns should also be raised about electromagnetic radiation at other frequencies.[19] X-rays and ultraviolet are examples of ionizing radiation from other parts of the electromagnetic spectrum and even visible light can be ionizing under some circumstances. However, electromagnetic radiation at lower frequencies does not have the energy to cause ionization, and microwaves and radio waves are both examples of non-ionizing radiation. In other words, people living near a radio transmitter are NOT being bombarded with ionizing radiation.

In spite of this, concerns have been expressed about potentially harmful effects caused by cell phones, cellular base stations, WiFi networks, microwave transmission links and satellite ground stations. Although these systems transmit at frequencies that cannot break atomic bonds, they can cause heating of the skin in much the same way as a microwave oven. To address this problem, the maximum output power has been restricted to levels at which the heating effect is far too small to cause cell damage.[20] However, concerns continue to be expressed about non-thermal mechanisms that might cause a risk to health. A study published in April 2008 found that exposure to mobile phone signals for just 5 minutes could stimulate human cells to divide; this process occurs naturally when tissue grows or rejuvenates, but it is also a key part in the development of cancer. It has also been suggested that a small percentage of the population suffers from "EMF sensitivity" and that this accounts for a wide range of problems, including headaches, tinnitus, nausea and stomach upsets.

Despite the lack of clear scientific evidence that using a mobile phone is a health hazard, it is reasonable to assume that holding a radio transmitter against the side of the head for long periods of time cannot be good for you. Powerful campaigns have been fought to prevent the construction of mobile phone masts near homes and schools after research suggested that children's brains absorb 50–70% more radiation from mobile handsets than adults because their skulls are smaller. Using similar arguments, WiFi networks in schools have been blamed for problems ranging from poor concentration in the classroom to Attention Deficit Hyperactivity Disorder (ADHD).

Such claims are easy to make and hard to refute. Poor concentration in the classroom could be blamed on a nearby WiFi network, but a diet of junk food combined with lack of exercise and late nights spent playing video games is a more likely explanation. People fall ill for all kinds of reasons and the fact that they used their mobile phone immediately before the first symptoms appeared does not prove cause and effect. In fact, epidemiological studies have so far failed to provide any convincing evidence that using WiFi networks or mobile phones leads to health problems.[21] However, we know that the health problems caused by exposure to ionizing radiation from a nuclear explosion may not appear until many years after the event, so it is possible that problems caused by using mobile phones or WiFi networks will not appear for a while yet. We need to remain vigilant until the evidence is completely clear. The probability of a long-term health risk is extremely small, but if such a risk were to be established, the impact could be devastating.

## 13.14 And finally . . .

Anyone who worked in the telecoms industry during the late 1990s will remember the excitement that was generated by the growth of competition and the emergence of the Internet as a mainstream communication medium. It was a time when anything seemed possible. Investors flocked to back start-up companies with astonishingly ambitious business plans, while established network operators were characterized as dinosaurs that were close to extinction. This, we were told, was a "paradigm shift". Anyone who counseled caution was told that they "just didn't get it".

It was a great party while it lasted, but it had to end eventually. When it did, it left behind a painful hangover that lingered for a number of years. Investors were spooked by the bursting of the dot.com bubble and telecoms companies suddenly found it hard to raise capital. Many of the shiny new networks that had been built during the boom were carrying little or no traffic, and bandwidth prices collapsed as network operators struggled for survival. Most of the smaller operators simply went bust or fell victim to the "dinosaurs". The value of telecoms assets collapsed, with many being sold off at firesale prices.

Since then, the industry has been slowly rebuilding itself. Most of the "irrational exuberance" has been replaced by the hard-headed realization that being in business is ultimately about making money. New technology may be exciting, but it is no longer seen as an end in itself. Telecoms today is a much more mature industry than it was in the late 1990s and it would be reasonable to expect that the same mistakes will not be repeated. Of course, the future will certainly bring challenges, but that's true of any large industry. On balance, the prediction is one of continued growth with rapid but manageable innovation. The biggest machine in the world will just keep getting bigger.

But is that enough? Are we now facing a future in which telecoms technologies are developed by big corporations in an orderly but unspectacular way? The inventions of the telegraph network, the telephone, the mobile phone and the Internet were all dramatic events that had far-reaching consequences. In fact, it is no exaggeration to say that the industry has produced some truly heroic figures who have single-handedly changed the world. Is it still possible for a talented individual to turn evolution into revolution, or have we seen the last of people such as Samuel Morse, Alexander Graham Bell, Guglielmo Marconi and Tim Berners-Lee?

Surprising as it may seem, there are still some major barriers that are holding back the march of telecommunications. For example, radio spectrum appears to be a finite resource that is set to become more and more congested as new mobile applications appear, and the speed of light is far too slow to meet the requirements of network designers.[22] It is certainly unfortunate that these barriers are believed to represent fundamental physical limitations, but truly great scientists and inventors are never afraid to challenge conventional thinking. Did Marconi give up when he was told that trans-Atlantic radio communication was impossible because radio waves could not bend to follow the curvature of the Earth? Network engineers have been achieving apparently impossible feats for more than 150 years and there is every reason to expect that the next generation of scientists and inventors will make their mark. We may not yet be able to see the next telecommunications revolution bearing down upon us, but that doesn't mean there isn't someone out there who can.

# Appendices

### Appendix A   Duplex telegraph

From the earliest days of telegraphy, it was possible to send and receive traffic alternately over a single wire. However, the "duplex telegraph" enabled a single wire to carry traffic in both directions *at the same time*. Figure A illustrates one of the methods used to achieve this apparently impossible feat.

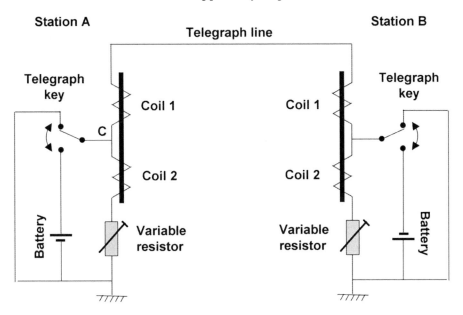

**Figure A.** Duplex telegraph.

For duplex telegraphy, we need to ensure that the sounder at Station B responds to the telegraph key at Station A (and vice versa). However, the trick is to prevent the sounder at Station A from also responding to the telegraph key at Station A (and vice versa) because this would interfere with the incoming message. This can be achieved by using a sounder with two separate coils.

When the telegraph key at Station A is pressed, point C is connected to the battery. The current flowing from the battery will split two ways at point C; some will head down through Coil 2 and return to the battery via the variable resistor, while the remainder will head up through Coil 1, across the telegraph line, through the telegraph key at Station B and back to the battery via the earth return. By adjusting the variable resistor at Station A, it is possible to ensure that the current splits equally between Coil 1 and Coil 2. Since the two coils have the same number of turns but are wound in opposite directions, the magnetic fields generated by the two coils will cancel out. This means that the sounder at Station A will not respond to the telegraph key at Station A. However, the current carried by the telegraph line does not pass through Coil 2 at Station B because it is shorted directly to earth by the telegraph key. This means that there is nothing to cancel out the magnetic field generated by Coil 1 and the sounder at Station B responds accordingly.

By symmetry, it can be seen that the same principle will operate in either direction (apart from the polarity of the battery, there is no difference between the equipment at Station A and Station B). Simultaneous bi-directional transmission is possible because the signals generated at the two ends of the line can be added together. If both keys are pressed simultaneously, then both sounders will respond.

## Appendix B  Baudot Code

A brief description of the Baudot telegraph system can be found in Chapter 2. This appendix describes the Baudot Code that was developed for use with the system, but which subsequently turned out to have a much wider range of applications.

The Baudot Code is a true digital code, with each character being represented by five bits. Each bit has only two possible values: 0 or 1. The 32 possible combinations of five bits are shown in the left-hand column of Table B-1.

**Table B-1.** Baudot Code.

| Code | Letter | Figure |
|------|--------|--------|
| 00000 | Blank | Blank |
| 00001 | E | 3 |
| 00010 | Line Feed | Line Feed |
| 00011 | A | - |
| 00100 | Space | Space |
| 00101 | S | ' |
| 00110 | I | 8 |
| 00111 | U | 7 |
| 01000 | CR | CR |
| 01001 | D | Enquiry |
| 01010 | R | 4 |
| 01011 | J | Bell |
| 01100 | N | , |
| 01101 | F | ! |
| 01110 | C | : |
| 01111 | K | ( |
| 10000 | T | 5 |
| 10001 | Z | + |
| 10010 | L | ) |
| 10011 | W | 2 |
| 10100 | H | £ |
| 10101 | Y | 6 |
| 10110 | P | 0 |
| 10111 | Q | 1 |
| 11000 | O | 9 |
| 11001 | B | ? |
| 11010 | G | & |
| 11011 | Figure Shift | Figure Shift |
| 11100 | M | . |
| 11101 | X | / |
| 11110 | V | ; |
| 11111 | Letter Shift | Letter Shift |

Notice that 32 separate codes are not sufficient to encode each of the 26 letters of the alphabet plus the digits 0 to 9 and assorted other characters. Baudot got around this problem by creating the Figure Shift (11011) and Letter Shift (11111) codes. When a Figure Shift code is transmitted, it means that the following codes should be interpreted using the Figure column in the table; similarly, if a Letter Shift code is sent, then the following codes should be interpreted using the Letter column.

Baudot's original code was devised in 1870 and was subsequently standardized as International Telegraph Alphabet No. 1 (ITA1). This code was modified in 1901 by a New Zealand sheep farmer called Donald Murray. After some further changes, it eventually became International Telegraph Alphabet No. 2 (ITA2) and this is the version of the code that is shown in the table. Baudot's choice of codes had been designed to make the code easy for operators to learn. However, Murray had developed a telegraph machine with a typewriter keyboard, so there was no longer any need for the operator to learn the codes. Instead, Murray's code was designed to minimize the number of bits that had to change when frequently used characters were transmitted, thereby reducing the wear and tear on the mechanical parts of his machine.

The ITA2 version of the Baudot Code was widely used for teleprinters and paper tape readers and punches until the 1960s, when its five-bit character set became increasingly restrictive. It could not, for example, handle lower-case letters. As computers started to appear, ITA2 was replaced by the seven-bit ASCII character set which persists (with some modifications) to this day. However, the ITA2 code has not completely disappeared—it is still used in specialist applications such as teletype communication between amateur radio enthusiasts and telecommunications devices for the deaf.

Surprisingly, the Baudot Code received considerable press coverage in August 2005 when it appeared on the cover of the *X&Y* album by British rock band Coldplay. Unfortunately, copyright restrictions have prevented reproduction of the album cover in this book, but a quick Internet search will show you what it looks like. Each vertical line on the album cover represents one code; the presence of a pair of colored boxes represents a 1 while a black space represents a 0. Once you know this, the image can be redrawn as shown in Table B-2.

**Table B-2.** Numerical representation of album cover.

| 1 | 1 | 0 | 1 |
|---|---|---|---|
| 0 | 1 | 0 | 0 |
| 1 | 0 | 0 | 1 |
| 1 | 1 | 1 | 0 |
| 1 | 1 | 1 | 1 |

Reading each column from the bottom upwards, we find that the first code is 11101. Using the table at the start of this appendix, we can see that this code represents X. The second code is 11011—the Figure Shift key—which indicates that the next code should be decoded as a Figure rather than a Letter. The following code is 11000, which represents 9. The final code is 10101, which decodes to 6. Coldplay presumably intended the diagram to represent X&Y, since this is the title of the album. However, a single bit error in the third character and the omission of a Letter Shift key before the final character mean that it actually decodes to "X96". A further coded message included within the album turns out to read "MAKE TRADE FAIR"—the name of an international organization that band members support.

## Appendix C  Microphone wars

At the Centennial Exhibition in 1876, Alexander Graham Bell exhibited two different types of microphone. One was a version of the electromagnetic microphone that he had used for most of his experiments. This design had the advantage that it was reasonably robust and no battery was required. However, the signal that it produced was weak, leading to problems on longer telephone lines. The alternative was the variable resistance microphone that had been used successfully in the famous "Mr Watson—come here—I want to see you" experiment. This form of microphone required a battery, which made it possible to generate a much more powerful signal. However, its use of a liquid created obvious practical difficulties for any telephone that was not clamped to a wall.

A year after the invention of the telephone, an order was placed for a system to link together five banks in Boston. When it was installed, the performance of the system was found to be unsatisfactory because the microphones used were not sensitive enough. The ideal solution to this problem would be a variable-resistance microphone that did not contain any form of liquid—such a microphone would be capable of generating a strong signal, but could be used safely in the working environment of a bank. By a stroke of good fortune, Bell was contacted at this moment by a young Bostonian called Francis Blake, who claimed to have invented just such a microphone. Even better, he was prepared to sell it for stock instead of cash—a blessing for Bell's cash-strapped company.

Blake's microphone was based upon the principal that the electrical resistance of some forms of carbon (such as coke, carbon granules or lamp black) varies in response to pressure changes. A diaphragm was used to detect the pressure fluctuations caused by the speaker's voice and the movements of the diaphragm were mechanically coupled to a capsule containing the carbon. A battery was used to drive an electric current through the carbon so that variations in the resistance of the carbon were translated into variations in the electric current. Within certain limits, increasing the battery voltage increased the strength of the electrical signal, thereby allowing the telephone to be used over longer distances. Since no form of electrical amplification had yet been invented, this was an important advantage.

It soon turned out that a number of other inventors had been working on improved microphone designs at about the same time and all of them had independently developed variants of the carbon microphone. The stage was set for a massive legal battle over patent rights.

The first of these inventors was Thomas Edison, who was later to find fame as the inventor of the phonograph and the electric light bulb. Edison's microphone had the additional refinement of a step-up transformer, which meant that current flowing through the carbon was smaller than the current transmitted down the line. Edison's work on the carbon microphone had been supported by Western Union, who had been seeking a way to circumvent the Bell patent.[1] However, Bell's company ultimately became the main beneficiary of Edison's work after their patent infringement suit against Western Union ended in an out-of-court settlement.

The next inventor of the carbon microphone was David Hughes, an Englishman who had spent the first 26 years of his life in the United States. Hughes was already wealthy as a result of his work on the printing telegraph. Rather than patenting his carbon microphone design, he effectively gave it to the world free of charge. This, of course, did nothing to endear him to Thomas Edison, who accused Hughes of plagiarism and patent infringement.

The third inventor of the carbon microphone was Henry Hunnings, an English curate. His patent was challenged by Edison and, lacking the financial resources to defend it, he sold the rights to Edison for £1,000.

The final inventor of the carbon microphone was a German-born American called Emil Berliner. Berliner had seen Bell's telephone at the Centennial Exhibition and had been inspired by what he saw. Berliner filed for a patent on the carbon microphone within 2 weeks of Edison, thereby laying the seeds for a long and bitter legal dispute. Berliner later went to work for Bell, who bought his design for $50,000.

All of these competing designs for the carbon microphone ultimately ended up under Bell's control. Later designs incorporated the best features of each of them and the carbon microphone became the standard telephone transmitter for over 100 years.

## Appendix D   Digital signal processing

Telephone networks often need to carry out some form of processing on the signals that they carry. The reasons for doing this might include:

- improving the quality of a signal by filtering out noise;
- eliminating echoes;
- detecting the presence of a signaling tone;
- determining whether a telephone circuit is carrying speech or some form of data (e.g. facsimile);
- inserting a call progress tone (e.g. dial tone, ringing tone, engaged, number unobtainable);
- setting up a conference call by bridging multiple telephone lines together.

In analog telephone networks, signal processing of this type is normally carried out using electronic circuits. Does this mean that digital networks have to convert signals back to analog whenever they need to process them?

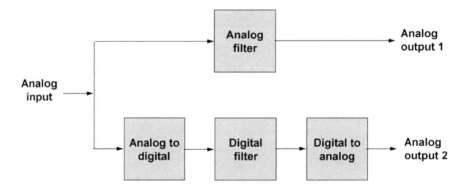

**Figure D-1.**   Comparison of analog and digital signal processing.

Fortunately, the answer to this question is No! In the top half of Figure D-1, an analog filter is used to process an analog signal. In the bottom half of Figure D-1, an equivalent digital filter is used to process the same signal. The analog filter is constructed using electronic components such as resistors and capacitors, while the digital filter is constructed using software running on a microprocessor; however, Analog Outputs 1 and 2 will be effectively the same. If a signal is converted to digital at the point at which it enters the telephone network and is not converted back to analog until it leaves the network, then digital signal processing can take place at any convenient point in the signal's path across the network. This appendix describes some of the ways in which digital signal processing is used in modern telephone networks.

Processing signals in their digital form can actually bring a number of advantages over the traditional analog approach. For example, digital systems do not drift as a result of temperature changes or component ageing. The performance of a digital system can be exactly simulated on a digital computer and production units are identical so no tuning is required. Since digital systems can be implemented in software, the same hardware can be reprogrammed to process signals in a number of different ways.

When you make a telephone call across a digital network, your speech is normally sampled 8,000 times per second. Each sample is then represented by 8 bits in the digital signal, so your telephone call requires a bit rate of:

$$8,000 \times 8 = 64,000 \text{ bit/second} = 64 \text{ Kbit/second}$$

In situations in which network bandwidth is congested, it may not be possible to allocate the full 64 Kbit/second to each voice channel, so further digital signal processing is required to reduce the bit rate. This process is known as "speech compression". Although there is inevitably a trade-off between bit rate and the subjective quality of the voice channel, it is often possible to reduce the bit rate by making changes that are not noticeable to a typical telephone user.

One of the first techniques of this type to be standardized for use in public networks goes by the rather catchy name of Adaptive Differential Pulse Code Modulation—or ADPCM for short. ADPCM can halve the bit rate of a standard digital voice channel with little or no impact on its subjective voice quality. As with most speech compression techniques, ADPCM attempts to remove redundant information from a speech signal. It does this by predicting the value of each incoming speech sample based on the speech samples that preceded it. It may not be immediately obvious that a speech signal is predictable in this way, but a little thought will show that voiced sounds are generated by the vibration of the vocal chords and any regular vibration is relatively predictable from one cycle to the next. Furthermore, telephone conversations involve periods of silence when the other person is speaking and it is reasonable to assume that a series of silent samples will be followed by another silent sample.

In ADPCM implementations, an identical predictor is used at the transmitter and at the receiver, as illustrated (in grossly simplified form) in Figure D-2.

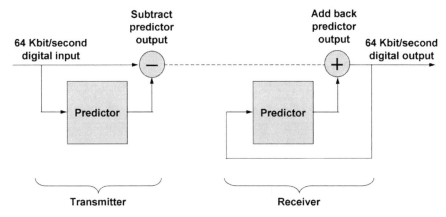

**Figure D-2.** Prediction in ADPCM.

Since both predictors are driven by the same sequence of inputs, their outputs are identical. As each speech sample arrives, the encoder compares the sample with the predicted value, and *it is the difference between these two values that is transmitted to the receiver.* So long as the predictor is reasonably accurate most of the time, the difference between each input value and the predicted value will usually be small. It therefore requires fewer bits to encode this difference accurately than it would do to encode the sample itself.

Many speech compression techniques exploit the fact that the human ear is relatively insensitive to quantization noise[1] if it is masked by a much larger speech signal. If a speaker is shouting, then a much higher level of quantization noise can be tolerated than would be the case if the speaker was whispering. It is not the absolute level of the quantization noise that counts, but its level in relation to the speech signal. We can therefore reduce the bit rate of a speech signal (and thereby increase the quantization noise) so long as we can ensure that the *ratio* of speech signal to quantization noise remains at an acceptable level.

Digital telephone networks have found a rather neat way of achieving this. Each speech sample is represented digitally by 8 bits, so the receiver output is restricted to 256 possible voltage levels.[2] If those voltage levels were to be evenly spaced, then the quantization noise would remain constant, irrespective of how loudly or softly the speaker was talking—the spacing between quantization levels is the same for large voltages (e.g. Point A in Figure D-3) as it is for small voltages (e.g. Point B).

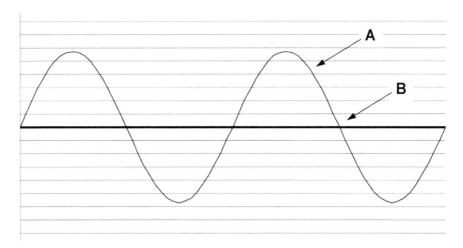

**Figure D-3.**  Evenly spaced quantization levels.

For clarity, not all of the 256 quantization levels are shown in this diagram. If, however, we were to distribute the quantization levels in a different way, then we could reduce the quantization noise for small signals at the expense of increasing it for larger signals. This is illustrated in Figure D-4.

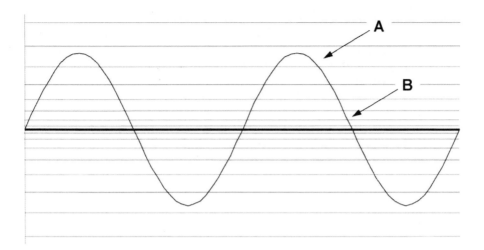

**Figure D-4.** Unevenly spaced quantization levels.

In this arrangement, the quantization noise at Point A will be larger than it was before, but it will still be masked by the strong speech signal. On the other hand, the quantization noise will be reduced at Point B where the speech signal is much weaker.

This illustrates the point that speech compression techniques are not always used to reduce the bit rate of a signal—in some cases, they can be used to improve the quality of a voice channel without changing its bit rate. This is done by improving the quality of the analog-to-digital conversion process (thereby increasing the bit rate) and then using digital signal processing to reduce the bit rate back to where it was before. The "non linear quantization" technique described above requires a more accurate analog-to-digital conversion to reflect the closely spaced quantization levels at Point B in Figure D-4, but it then sacrifices some of this accuracy for larger signals in order to constrain each sample to 8 bits.

The speech compression techniques that have been described so far belong to a class of speech coding techniques known as "Waveform Coders". Waveform Coders aim to produce an output waveform that matches the input waveform as closely as possible. Some Waveform Coders can transmit acceptable quality speech at bit rates as low as 8 Kbit/second, but bit rates below about 16 Kbit/second are usually handled better by another class of speech coding techniques called Source Coders.

Source Coders work on an entirely different principle from Waveform Coders. Their aim is to produce a speech signal that is perceived to have good quality— even if the output waveform does not accurately match the input waveform. Source coding techniques attempt to build a model of the speech production process. For example, voiced sounds (such as "r") are created by the vibration of the vocal chords and by the way in which this sound is subsequently modified by factors such as the shape of the mouth and the position of the tongue. Unvoiced sounds (such as "f") are created by forcing air past the tongue, lips and teeth, and do not involve the vocal chords. A Source Coder attempts to model these processes and it is the parameters required to drive this model that are transmitted down the line.[3]

Source Coders can achieve intelligible speech at bit rates as low as 1 Kbit/second and this raises an interesting question: if a standard telephone channel occupies 64 Kbit/second, while a Source Coder can reduce this to 1 Kbit/second, then why are Source Coders so rarely used in standard telephone networks?

The answer, sadly, is that you don't get something for nothing in this world. All speech compression techniques have to make compromises in one form or another and these compromises become more obvious at lower bit rates. Very-low-bit-rate voice channels tend to sound synthetic and speaker recognition is often a problem at the receiving end. Furthermore, the performance of low-bit-rate voice channels for non-speech signals is often very poor, even if they sound fine for voice. This is a problem for dial-up modem traffic, which is unlikely to survive any encounter with a low-bit-rate voice channel.

For these and other reasons, speech compression is not widely used in traditional telephone networks. In the past, speech compression had a role on very expensive international links (such as trans-Atlantic satellite links), but dramatic advances in fiber optics (see Chapter 6) have driven down the cost of bandwidth and undermined the justification for speech compression on many of these routes. However, just as it seemed that technology advances might eliminate the need for speech compression, the requirement to carry voice signals across packet-switched networks created a whole new set of applications.

The speech coding technologies that we have discussed so far were designed for fixed bit rate situations (i.e. the channel operates at the same bit rate irrespective of whether anyone is speaking or not). However, packet-based networks (such as the Internet) are capable of providing bandwidth on demand. It is therefore very tempting to reduce the average bit rate required to carry a telephone conversation by transmitting speech packets only when somebody is actually talking. If we assume that a telephone conversation at any given moment consists of one person speaking and the other person listening (there are exceptions to this rule, of course!), then one direction of transmission will always be carrying silence. If we only transmit speech packets when someone is actually talking, then the average bandwidth required to carry a telephone conversation can be reduced by at least a factor of two.[4]

Although the simplicity of this technique is appealing, it has to be used with care. Strange as it may seem, problems can be caused if the output of the receiver goes completely silent whenever packets stop being received. If, for example, the speaker is talking in a noisy environment (such as on a train), then the listener will hear the background noise cutting in during periods of speech and then cutting out again. This can be highly disconcerting. In fact, people will even complain if they cannot hear the low-level hiss that they are accustomed to during quiet moments in a telephone conversation. For this reason, some speech coding techniques introduce artificial background noise during periods of silence.

Another problem with speech compression techniques is the amount of additional transmission delay that they introduce. Although this is not a problem for ADPCM, many lower-bit-rate coding techniques need to buffer a significant number of speech samples before they can calculate their next output and this delays the transmission of the speech. The problem is made worse in the context of packet-switched networks because each packet has to hang around until the transmitter has filled it up with data. The packet then encounters further delay as it crosses the network and this delay is not constant so buffering is required at the receiver to eliminate the "jitter". This combination of delays can mean that a word will not emerge from the receiver until 100 milliseconds[5] or more after it was spoken. This may not sound serious, but it is quite long enough to cause very distracting echoes on a telephone call.

Most of us have encountered the problem of echo on long-distance telephone calls at one time or another. Echoes typically occur when the far end of the line is not properly balanced, so your speech is reflected back towards you. The problem normally arises at points at which a conversion takes place between "2-wire" and "4-wire" circuits, as illustrated in Figure D-5.

**Figure D-5.** 2-wire to 4-wire conversion.

The telephone line to your home is likely to use a single pair of copper wires to carry speech in both directions and so would be referred to as a 2-wire circuit. In contrast to this, the trunk connections between telephone exchanges use a separate pair of wires for each direction of transmission and so are referred to as 4-wire circuits.[6] Trunk connections use 4-wire circuits because it makes it simpler to introduce amplification or digital regeneration at intermediate points along the route. Connections from subscribers to the local exchange are too short to require amplification, so 2-wire circuits are used to reduce costs.

In Figure D-5, Sam's voice is carried up the telephone line from his home to the 2–4 wire converter in the local exchange. It is then carried on 4-wire circuits until it reaches Louise's local exchange, at which point it is converted back to 2-wire to suit Louise's telephone line. However, if the 2–4 wire converter at Louise's local exchange is not properly balanced, then it is possible that some of the speech will leak through the converter from Point D to Point F, and will be returned to Sam as an echo. If this speech takes (say) 100 milliseconds to reach the far end of the line, and a further 100 milliseconds to come back again, then the echo will be heard after a delay of 200 milliseconds.

A simple solution to this problem is to use a device called an echo suppressor, which ensures that only one direction of transmission is available at any moment in time. The principle is illustrated in Figure D-6.

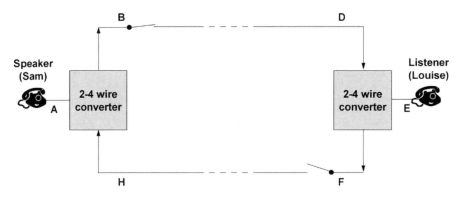

**Figure D-6.** Echo suppressor.

As soon as Sam begins to speak, the switch at Point B closes so that his speech is transmitted to the far end of the line and the switch at Point F is opened so that no echo can return. Whenever Louise is speaking, the settings of these switches are reversed. In the past, this technique was used extensively on analog trans-Atlantic links, but the switchover that occurs when the listener starts to speak is sometimes too slow to catch the opening syllable and this can lead to "choppy" speech. Problems also occur when both people start to speak at the same time.

These difficulties can be avoided by using echo cancellation instead of echo suppression. Echo cancellers use digital signal processing to generate a synthetic version of the echo; this synthetic echo is then subtracted from the incoming signal, thereby removing the real echo. The process is illustrated in Figure D-7.

Sam's speech will pass through Points A, B, C, D and E before arriving at Louise's telephone. As before, a proportion of this signal will leak through from Point D to Point F, and will head back up the line towards Sam. However, the echo canceller at Louise's end of the line will have generated a "synthetic echo" that matches the real echo as closely as possible. At Point G, the synthetic echo is subtracted from the signal arriving from Point F, thereby sending an echo-free signal to Point H.

If Louise now starts to speak, her speech will arrive at Point G via Point F. However, the echo canceller at Louise's end of the line will not have seen

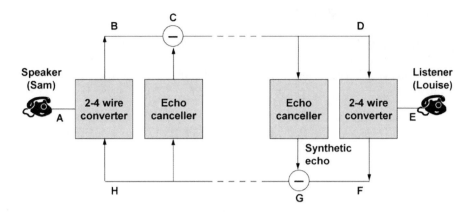

**Figure D-7.** Echo cancellation.

Louise's speech signal at its input and so will not generate a synthetic echo to cancel it. As a result, Louise's voice will be transmitted up the line unchanged. If any echoes occur at Sam's end of the line, then they will be dealt with by the echo canceller at that end of the line.

Unlike an echo suppressor, an echo canceller does not disconnect either direction of transmission, so speech "choppiness" is not a problem. Furthermore, both callers can talk simultaneously without either of them hearing echoes. As a result, echo cancellers are widely used on long-distance trunk circuits where the delay is long enough to create a risk of echoes[7] and in situations in which speech is transmitted across packet switched networks.

## Appendix E  DSL technologies

Asymmetric Digital Subscriber Line (ADSL) is a technology that allows a single twisted pair telephone line to support broadband Internet access whilst simultaneously carrying telephone calls. ADSL uses Frequency Division Multiplexing[1] to separate the telephony signals from the data traffic, as illustrated in Figure E-1.

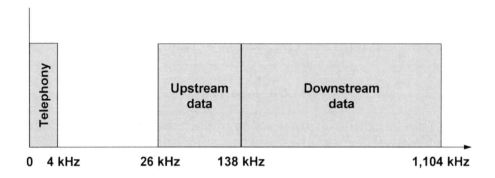

**Figure E-1.**  ADSL frequencies.

As explained in Chapter 5, a telephone conversation is restricted to frequencies in the range 300 Hz to 3.4 kHz. ADSL uses frequencies in the range 26–138 kHz to carry upstream data (going to the Internet) and a much larger band from 138 kHz to 1,104 kHz to carry downstream data (coming from the Internet). As a result, the bandwidth provided from the network to the user is considerably greater than the bandwidth provided in the opposite direction. This improves the performance of the service for web surfing—but not for file sharing and other activities requiring a fast uplink. It is this asymmetry that puts the "A" in ADSL.

Figure E-2 illustrates how broadband services are delivered over a twisted pair telephone line using ADSL.

A device called a splitter[2] is used at each end of the telephone line to separate the low-frequency signals carrying the analog telephony from the higher frequencies that carry the Internet traffic. As a result, the telephone at the bottom of Figure E-2 is connected across to the switch at the local exchange (as it would be on a telephone line without broadband), while the ADSL Modem & Router is connected to a device called a Digital Subscriber Line Access Multiplexer (DSLAM). Since the Internet traffic is carried by frequencies that are well above the range of a telephone call, it is possible to chat on the telephone and surf the web at the same time.

In order to use the broadband service, each customer requires an ADSL modem. If, as illustrated in Figure E-2, they want to connect more than one device to the Internet, a router is also required—in the example shown, the customer is using a VoIP adapter to make telephone calls over the Internet and is

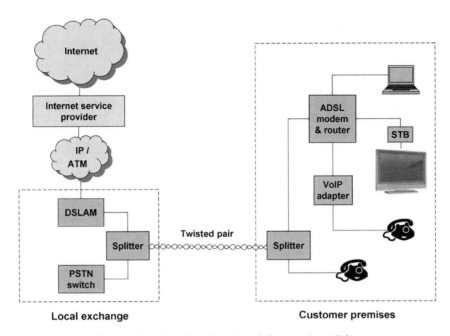

**Figure E-2.** Broadband service delivery using ADSL.

also using a set-top box (STB) to receive IP television services. The traffic from the customer's ADSL Modem & Router is received by the DSLAM at the local exchange, from where it is forwarded to the Internet.

ADSL is one member of a family of technologies that provide digital services over twisted pair telephone lines. In addition to ADSL, this family also includes HDSL, VDSL and SHDSL.

BT has used HDSL to deliver 2 Mbit/second digital services to customer sites using existing copper cables. However, HDSL can require up to three copper pairs to deliver a single digital service (depending upon the length of the copper cable and the bandwidth required) and there are sometimes interference issues with other digital services carried by different pairs in a multi-core cable.

In 2001, Single-pair High-speed DSL (SHDSL) was ratified as a standard. It allows data rates up to 2.3 Mbit/second to be delivered over a single copper pair and SHDSL and ADSL services can be carried together in the same cable without interference.

VDSL can achieve data rates of over 50 Mbit/second over short distances and was developed to support extremely high-bandwidth applications such as high-definition television (HDTV). However, a typical telephone line is too long to support such high data rates, so VDSL services have to be delivered from a street cabinet close to the customer's home. VDSL2 is a recent enhancement to the standard that provides improved performance. ADSL has also been enhanced with the appearance of new and improved variants such as ADSL2+.

## Appendix F  Leveling up the playing field

The huge cost of building an access network based upon copper wire explains why very few network operators have ever attempted to do it. Most Western European countries developed their copper access networks at a time at which telecoms was a state-owned monopoly, so the costs were borne by the long-suffering tax payer. When telecoms markets were opened up to competition in the 1980s and 1990s, access networks became the property of privatized former monopolies such as BT, France Telecom and Deutsche Telekom. This created rather lop-sided markets, because the company that owned the access network was inevitably more equal than its competitors. Since it was not economically viable for competing network operators to build their own access networks,[1] governments and regulators looked for ways to address this imbalance.

One of their early initiatives was a scheme called indirect access. Indirect access forces the company that owns the access network to make it available for use by competing operators on fair and non-discriminatory terms. In the United Kingdom, for example, BT had to offer interconnect arrangements to operators such as Cable & Wireless (C&W). This meant that C&W could terminate a call by handing it over to BT at an interconnect point between the two networks and BT would then deliver it using its access network. Similarly, C&W customers with a

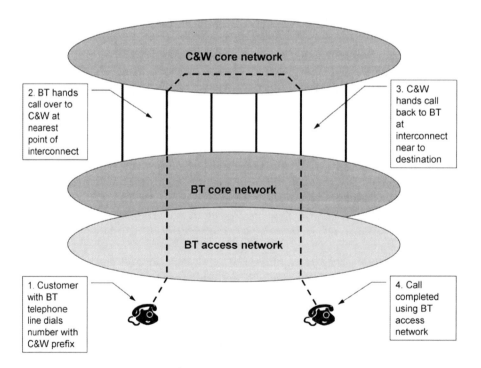

**Figure F.**  Indirect access.

BT telephone line could make calls over the C&W network by dialing a short numerical prefix in front of the telephone number. Figure F should help to explain things.

This may seem like a rather convoluted way of handling a telephone call and it certainly doesn't make much sense for local calls. However, for national calls, the C&W network is providing the long-haul part of the link, so it is adding significant value. This arrangement allowed a number of companies to offer telephone services in direct competition to BT without the crippling expense of building and operating an access network. In many cases, the customer was unaware that their calls were being handled in this way; they received a bill from BT for their line rental and a bill from C&W (or whoever) for the cost of their calls. All the other financial settlements relating to these calls took place directly between the network operators.

Indirect access did create a degree of competition in the telephony market, but it did nothing to stimulate competition in other forms of service. As the Internet became increasingly important, the need arose to find a way of promoting competition in broadband services. This time, a regulatory requirement called local loop unbundling was used. Local loop unbundling enables a customer to ask for their telephone line to be taken over by a service provider of their choice. The service provider installs their own equipment at the BT local exchange and uses that equipment to deliver services to the customer. These services would typically include broadband Internet access, but might extend to telephony services and even to digital television. BT retains ownership of the copper wire and is still responsible for repairing any faults that arise on the line.

## Appendix G   Fixed wireless access networks

Fixed wireless access systems fall into two distinct categories: point-to-point and point-to-multipoint. In a point-to-point system, the transmitter produces a focused beam of microwaves that is carefully aimed at the corresponding receiver. This maximizes the amount of transmitted energy that arrives at the receiver while minimizing the interference caused to other nearby radio systems. In contrast to this, a point-to-multipoint system uses a single base station to flood an area with coverage; each user of the system communicates with the same base station, so the capacity of the system is shared by all the users in the area. The differences between the two systems are illustrated in Figure G-1.

**Point-to-point**

**Point-to-multipoint**

**Figure G-1**.  Point-to-point and point-to-multipoint networks.

While point-to-point systems can deliver large amounts of bandwidth to a single user, point-to-multipoint systems are better at providing smaller amounts of bandwidth—or short bursts of high bandwidth—to a larger number of users. As an example, "WiFi" wireless networks are often used to provide point-to-multipoint high-speed Internet access in homes and offices.[1] In each case, a base station—or "access point" in WiFi parlance—communicates with a number of computers equipped with wireless interfaces. The access point typically has a broadband connection to the Internet that is shared by all the users on the WiFi network.

The success of WiFi has been driven by a number of factors. The technology is based upon open standards with catchy names like IEEE 802.11g, so a wide range of manufacturers are producing compatible WiFi equipment and prices have been driven down as a result. Since wireless networks do not require cabling, they can be installed quickly and cheaply. Furthermore, WiFi operates in unlicensed radio bands, so there is no need to apply for a license. Of course, there is always the risk of interference between wireless networks in adjacent buildings, but this can usually be cured by switching to a different frequency channel. Even if interference does occur, it simply reduces the available capacity of the network—it does not cause messages to go astray.

WiFi "hotspots" are often found in places such as hotels, coffee shops, railway stations and airport departure lounges. More recently, we have seen initiatives to provide city centers with WiFi coverage using a network of WiFi base stations. However, WiFi was originally designed to provide coverage within a single building and it is less suitable for providing coverage across a metropolitan area. For this reason, a new family of wireless standards known as WiMAX (or Worldwide Interoperability for Microwave Access) have been developed. The first version of WiMAX to be widely implemented[2] is designed to provide fixed wireless access in rural areas where wired broadband services are not available or in countries where there is little or no fixed infrastructure. It could, in theory, compete directly with existing broadband services, although the evidence so far suggests that it will struggle to do this. Another version of the WiMAX standard[3] provides coverage for mobile users in cars or trains and this version has the potential to compete with 3G mobile networks in some situations.[4]

WiMAX will typically cover a much larger area than WiFi, although coverage is heavily dependent upon whether a clear line of sight is available between the base station and the user.[5] Systems that do not require line of sight are restricted to shorter distances but they have considerable cost advantages. Instead of engineers having to visit the house to install an aerial, the receiving equipment can be mailed to the customer through the post. When it arrives, the customer simply plugs the device into the mains, connects it to the PC *et voilà*—instant broadband!

An alternative to WiMAX's point-to-multipoint approach is to use a self-configuring mesh radio system. Mesh radio systems avoid the need for a base station that has a line-of-sight connection to every radio in the area. Instead, each radio can act as a base station to allow other radios to join the network. A radio connects to a nearby radio, which, in turn, connects to another radio until a mesh of interconnections has been created. This concept is illustrated in Figure G-2.

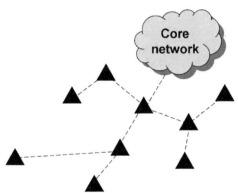

**Figure G-2.** Mesh radio network.

Mesh radio has a number of attractive features. While a tall building might be a major obstacle for a point-to-multipoint network, a mesh network can find a way around it using a series of "hops". Furthermore, if a new obstacle subsequently appears, or if one of the radios fails, then the network will automatically reconfigure itself to fix the problem. As a result, mesh radio systems can achieve almost 100% coverage and high reliability. In order to provide compatibility between equipment from different vendors, work is currently under way to develop a standard[6] based upon WiFi.

Mesh radio systems have been used to provide WiFi services across metropolitan areas, but attempts to use mesh radio to provide broadband services to rural homes located beyond the reach of ADSL have been less successful. BT conducted a trial in 2002, but it appears that they encountered operational issues and the system was not cost-effective. The equipment supplier subsequently went out of business.

As explained in Chapter 7, fixed wireless access has not proved to be particularly successful for delivering telephony services to residential customers. However, point-to-point radio does have a role in providing services to business customers. Using radio is often cheaper than digging up the street to install a cable—particularly in city-center locations or in areas where natural obstacles like rivers make cabled solutions prohibitively expensive. The costs of the radio link are justified because business customers tend to have far higher telephone bills than residential customers. Furthermore, radio links can be installed quickly so the customer is given less time to change their mind! If a cable is subsequently installed, the radio link can be retained as a back-up path or the equipment can be re-used to serve another customer.

## Appendix H    Internet Service Provider networks

The Internet is made up of a large number of interconnected networks that follow a common set of technical standards. Interconnection between these networks can take one of two forms, depending upon whether or not any money changes hands. A peering relationship is, as the name suggests, a meeting of equals. In a peering relationship, the interconnection is considered to be mutually beneficial for both parties and so no charges are levied by either party when traffic is sent across the interface. In contrast to this, a transit relationship is used when one network is significantly bigger and better connected than the other. In this case, the relationship is likely to benefit the smaller network much more than the larger network, so the larger network will levy charges for the transit service that it is providing.

Internet Service Provider (ISP) networks fall into three distinct groups:

1. *Tier 1*. These networks form the core of the Internet. To qualify as Tier 1, an ISP network must have a peering relationship with every other Tier 1 network on the Internet.
2. *Tier 2*. These networks have a peering relationship with one or more Tier 1 networks, but not with all of them. As a result, they still have to purchase transit services from one or more Tier 1 networks to reach some parts of the Internet.
3. *Tier 3*. These networks have no peering relationships with any Tier 1 network. They obtain all their Internet connectivity by purchasing transit services from Tier 1 or Tier 2 networks.

All of these networks—irrespective of their size and status—conform to the same addressing structure. A host computer that is permanently connected to the Internet will have a unique IP address that is recognized on all Internet networks.

## Appendix I  The Internet address shortage

One of the functions of the IP protocol is to support a global addressing structure that allows a host computer on one network to communicate with a host on another network. Every device connected to the Internet has a unique IP address, in the same way as every telephone in the world has a unique telephone number.

An IP address consists of four separate numbers in the range 0 to 255. They normally look something like this:

123.45.67.8.

As a general rule, the digits at the start of the address identify the network, while the digits at the end of the address represent an individual host connected to that network. For example, if a particular network has been allocated 256 addresses, then they might take the form:

123.45.67.0
123.45.67.1
:
:
123.45.67.255.

In this case, the first three numbers identify the network, while the final number represents the individual host.

In order to ensure that every host connected to the Internet has a unique address, blocks of IP addresses are allocated and managed centrally.[1] However, the management of these addresses within a particular network is the responsibility of the network operator. In the example above, the block of addresses beginning 123.45.67 would be allocated centrally, but it is up to the network operator to decide which host is given the address 123.45.67.8.

Mathematically minded readers may have concluded from this that the Internet must be running short of addresses. If each address contains four numbers and each number can have 256 different values (including zero), then the maximum number of possible addresses is:

$256 \times 256 \times 256 \times 256 = 4{,}294{,}967{,}296.$

In the days when Kahn and Cerf were developing the IP protocol, this must have seemed like a pretty large number. However, the prodigious growth of the Internet, coupled with the migration from mainframes to personal computing, means that we are likely to run out of spare Internet addresses within a few years. The most obvious solution to this problem is to define a new Internet address format that can accommodate a much larger number of addresses and this has been done: the latest version of the IP protocol[2] supports a mind-boggling $2^{128}$ addresses. To illustrate just how huge this number is, it represents approximately (deep breath) 50,000,000,000,000,000,000,000,000,000,000 addresses for every man, woman and child alive today! Whilst this will certainly solve the problem, upgrading the whole of the Internet to this new standard will not be a simple task. To meet the immediate need, alternative address conservation techniques are required.

Dynamic address allocation is one technique that can be used by ISPs to reduce the number of IP addresses that they need to allocate to users. The technique is based on the assumption that at any given moment in time, the number of *active* users on their networks is very much lower than the number of *registered* users. This technique worked particularly well in the days of dial-up Internet access because users only connected to the network for relatively short periods of time. When a connection was established, the user would be allocated an address from a pool of available addresses and this address would be returned to the pool at the end of the user's session. Although dynamic address allocation is also used for broadband access to the Internet, the always-on nature of broadband services means that each user requires an address for much longer periods of time.

Network address translation (NAT) is another technique that is widely used to conserve IP addresses. The basic idea here is that an address that has been allocated to one network can be re-used on another network *so long as nobody on the Internet can see it*. A simple example of this technique is often found in wireless (WiFi) home networks, where a single broadband Internet access service is shared by a number of separate computers located in different parts of the house. In these home networks, the computers are linked to the Internet via a small wireless router. When the router is turned on, the ISP issues it with an address using dynamic address allocation. This is known as a "public" address, because it is visible to all the other machines on the Internet. The router, in turn, allocates addresses to each of the computers on the home network. However, the addresses allocated by the router are "private" addresses; they are not visible on the Internet and so it does not matter if they are also being used by someone else. When one of the machines on the home network wishes to send a message to a machine somewhere else on the Internet, it sends a packet over the wireless link to the router. The router removes the computer's private address from the packet header and inserts its own public address before launching the packet into the Internet. So far as the rest of the Internet is concerned, that packet has been generated by the router. Any reply would be sent back to the router, which would then forward it to the machine that generated the original message.[3]

Each computer on the wireless home network is effectively invisible to hosts out on the Internet. This means that Internet hosts cannot send messages to computers on the wireless network unless they are responding to a request—any unsolicited packets will simply be discarded by the router. This can provide valuable protection against Internet worms and other nasties, but it can also prevent the correct operation of certain Internet protocols.

### Appendix J  Virtual private networks

Let's imagine a fictitious company based in Bristol. They have 10 employees and the computers in their office are linked together by a Local Area Network (LAN). The office network provides access to the Internet and there are several servers on the network to provide services such as file storage and printing. This situation is illustrated in Figure J-1.

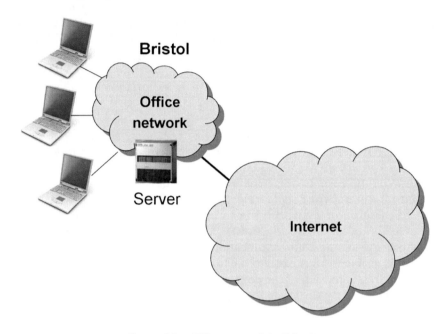

**Figure J-1.**  Office network in Bristol.

One of the servers on the office network hosts a small "intranet" (the "intra" rather than "inter" in the name indicates that an intranet is essentially a private version of the Internet). The intranet is used to improve business efficiency; it provides product data, recent press releases, an internal telephone directory and access to the timesheet and expenses systems. An intranet is built using Internet technologies and employees can access intranet sites using an ordinary web browser. However, intranet sites are not visible to other users on the Internet.

Now let's suppose that the company opens a second office in Bath. The employees located in the Bath office need access to the Internet, but they also need access to the corporate intranet and to some of the other services hosted on the servers in Bristol. One way of achieving this is illustrated in Figure J-2.

The networks in the Bristol and Bath offices are linked together by a dedicated digital connection called a leased line. Users in the Bath office can gain access to both the intranet and the Internet via the Bristol office.

Now let's imagine that the company employs someone who regularly works

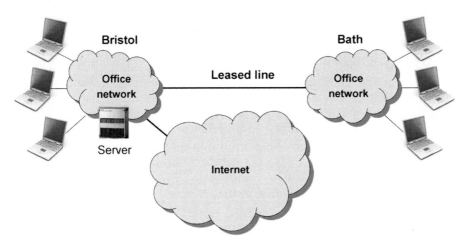

**Figure J-2.**  Linked office networks in Bristol and Bath.

from home. The company could, in theory, rent another leased line to provide a link to that person's home, as illustrated in Figure J-3.

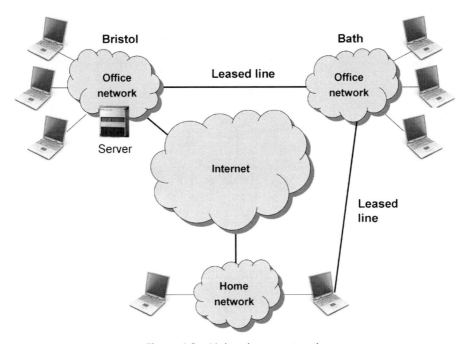

**Figure J-3.**  Link to home network.

This solution has a number of weaknesses. To begin with, it could start to get messy as the number of leased lines increases and the network starts to resemble a pile of spaghetti. Second, leased lines are usually expensive and they tend to get more expensive as the distance increases; Bristol and Bath are not far apart, but the costs could rise significantly if the company opened a new office in Scotland or overseas. Finally, leased lines do not provide a solution for employees who are required to travel.

If the salesman's laptop is connected to the Internet, then it should be possible to make a connection via the Internet to the office network in Bristol and this would provide a much cheaper solution than a leased line. However, the Internet is notoriously insecure, so the company would naturally be concerned about using it to carry sensitive commercial information. The solution, as illustrated in Figure J-4, is to create a Virtual Private Network (or VPN for short).

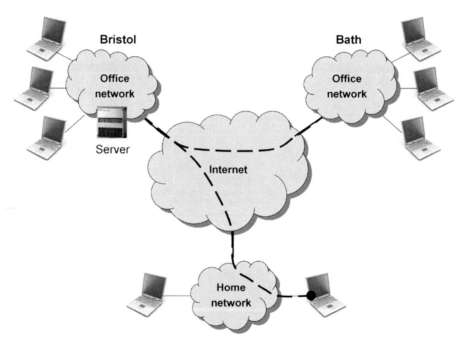

**Figure J-4.** Virtual Private Network.

Notice that the leased lines are no longer required and the connection between the two office networks is now provided by a secure VPN "tunnel" across the Internet. A packet being sent from Bath to Bristol will be intercepted before it leaves the Bath office network, the entire packet (including the header) will be encrypted and the result will be placed in the payload of another IP packet for transmission across the Internet. At the Bristol office, the payload is decrypted to reveal the original IP packet, which can then be delivered to its

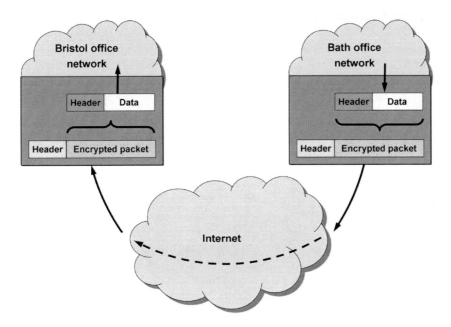

**Figure J-5.** Transmitting a packet from Bath to Bristol.

intended destination on the Bristol network.[1] This sequence of operations is illustrated in Figure J-5.

This idea of hiding one packet inside another is a useful trick and one that is widely used. The tunnel does not prevent packets from being intercepted as they make their way across the Internet but, thanks to the encryption, anyone intercepting a packet would find nothing intelligible in the payload. Even the packet header provides little useful information—the source and destination addresses are simply those of the VPN devices at either end of the tunnel, so they reveal nothing about the internal structure of the two office networks.[2]

For security reasons, the office networks in Bristol and Bath are both connected to the Internet using a device called a firewall. Each firewall restricts the types of traffic that can pass between the office network and the Internet, thereby protecting the office network against various types of Internet threat.[3] Since some form of VPN device is required at each end of a tunnel to perform the encrypt/decrypt operations described above, it is often convenient to implement VPN functionality within the firewall. In the case of the home network shown in Figure J-4, the VPN tunnel would probably be terminated by software running on the user's laptop.

## Appendix K   Internet voice services

In a conventional telephone network, voice calls are set up and cleared down by telephone exchange switches. In an IP network such as the Internet, circuit switches are replaced by routers and "softswitches", as illustrated in Figure K-1.

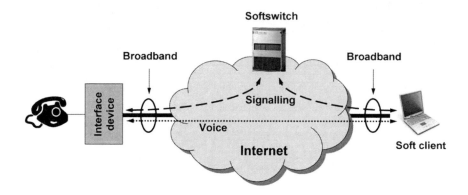

**Figure K-1.**   Telephone call across the Internet.

This diagram illustrates a telephone call across the Internet. The call originates on a conventional telephone connected to the Internet via a broadband interface device and terminates on a soft client running on a PC. The softswitch provides the intelligence needed to set up and tear down the call and it captures the data required to generate a phone bill. In a conventional telephone network, this intelligence would be found in a telephone exchange. However, in contrast to a conventional telephone exchange, the softswitch does not carry any voice traffic—once the two ends of the line know how to send packets to each other across the Internet, the softswitch can let them get on with it. As a result, the softswitch can be physically located anywhere on the Internet without having any significant effect on the quality of the call.

A softswitch runs on a standard computer so the hardware is relatively cheap. The software, of course, is rather more expensive, but the price is often based on the maximum number of calls that the softswitch can support, so it is usually much more cost-effective to build a small voice network using softswitches rather than conventional telephone exchanges. Since the IP packets are directed to their destination by the routers in the IP network, there is no need to buy any additional hardware to switch voice calls across the network. Furthermore, it is possible to use digital speech compression techniques to reduce the bandwidth allocated to each voice channel and further bit rate reductions are possible by not sending packets during periods of silence in the conversation. All of these factors help to reduce the cost of Internet telephony.

There are many people in the world who have access to a telephone (either fixed or mobile) but do not have broadband access to the Internet. For this reason, Internet voice services typically allow calls to originate and/or terminate on conventional telephone networks. These services use a device called a PSTN[1] gateway to link the packet-based world of Internet telephony with the circuit-switched world of the PSTN, as illustrated in Figure K-2.

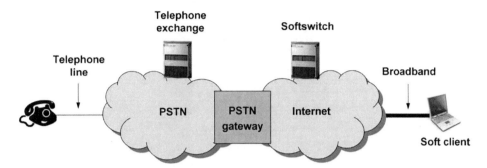

**Figure K-2.** PSTN gateway.

The telephone network has been optimized to carry voice traffic, while IP was designed primarily to be a data protocol. An IP network will quite happily drop the occasional packet or delver packets out of sequence. Although the TCP protocol can be used to sort out these problems by requesting the re-transmission of lost or damaged packets, this adds delay—and delay is one impairment that voice cannot tolerate. Imagine a conversation in which there is an occasional unexpected silence followed by words arriving at several times their normal speed! The problem of packets arriving at an uneven rate can be smoothed out by adding buffering at the receiver, but this makes the overall delay even longer because every packet is now being delayed to match the slowest. As a general rule, increasing the end-to-end delay makes conversation more difficult because the participants keep interrupting each other—a problem regularly encountered on satellite links.

In order to minimize delay, voice services over the Internet do not make any attempt to re-transmit lost or corrupted packets;[2] instead, the lost information is simply replaced with a period of silence or the gap is filled with some form of background noise to make it less noticeable. Unfortunately, however, packet re-transmissions are not the only source of delay on the Internet; if the network is congested, then voice packets can find themselves stuck at the back of a queue of packets waiting to be transmitted over a link. Various schemes have been devised to allow voice packets to jump to the front of any queues that they encounter but the laissez-faire culture of the Internet means that end-to-end quality cannot be guaranteed for every call.

In order to address this problem, network operators have started building quality-enabled IP networks that are not part of the Internet. If one organization has control over a network, then quality standards can be enforced on that network and it can be used to offer a full range of voice and data services. IP networks of this type are starting to appear in the core of existing telephone networks and—although you may not be aware of it—some of your telephone calls may already be passing over them. These "next generation" IP networks are discussed in Chapter 13.

## Appendix L   IP television

The bandwidth required to transmit standard-definition television is about 1.5 Mbit/second, even if modern compression technologies are used. This bandwidth rises to about 7.5 Mbit/second for high-definition television. Most standard broadband services cannot deliver such high bandwidths on a continuous basis, but it is now possible to deliver "IPTV" television services over copper telephone lines using high-speed broadband networks that have been optimized for this purpose. This appendix discusses some of the techniques used to deliver broadcast television services and Video on Demand over an IP network.

Video on Demand (VoD) services allow users to view content of their choice whenever they want—usually after payment of a fee. It is a bit like having access to a DVD player with a huge library of DVDs. A typical VoD service is illustrated in Figure L-1.

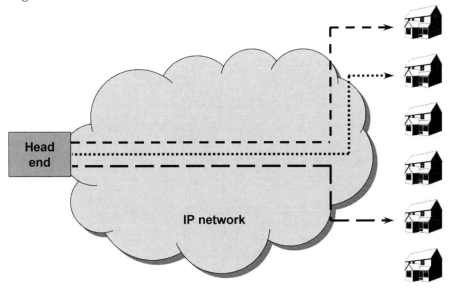

**Figure L-1.**  Video on Demand.

The content (movies, TV programs, music videos, etc.) is stored on a video server located at a "head end". If a user selects a particular movie and presses the Play button, the video server will stream that movie across the IP network to a set-top box located in the user's home. The set-top box strips away the IP packets and presents the video signal in a form that the TV can display.

Now let's suppose that several people living in the same street decide to watch a movie at the same time. Even if they decide to watch the same movie, a separate video stream will be created for each user. However, our earlier calculations have indicated that each video stream is likely to need a fairly meaty chunk of bandwidth. If a large number of people all decide to use the VoD service at the same time, the network could literally run out of bandwidth.

One way to address this issue is by introducing a device called a cache, as shown in Figure L-2.

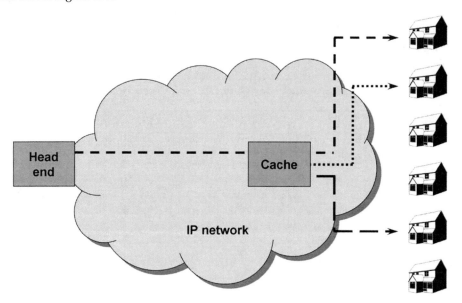

**Figure L-2.** Video cache.

The first time that a movie is viewed, it will be streamed from the head end as before. However, as it gets close to the viewer's home, it passes through the cache, which stores a copy of the movie in its memory. If someone else subsequently requests the same movie, it will be streamed from the cache rather than from the head end, thereby reducing the load on the IP network. In this way, popular content tends to be delivered from the cache rather than the head end. Any content that has not been viewed for a while will be deleted from the cache to make room for something else.

Of course, IPTV services don't just deliver Video on Demand—they can also carry broadcast television channels (such as BBC1 or CNN). However, the problems encountered when carrying broadcast TV are quite different from the problems of Video on Demand. The basic principles are illustrated in Figure L-3.

Only one video stream is required to broadcast a TV channel across the IP network if a technique called "multicasting" is used to split the stream across all the households that wish to view it. A multicasting router makes a number of copies of each incoming video packet and then sends each copy down a different branch of the network. These copies, in turn, are likely to encounter additional multicasting routers as they make their way across the network, with further copies being created as a result. This process means that each video packet leaving the head end is duplicated repeatedly until there are enough copies to

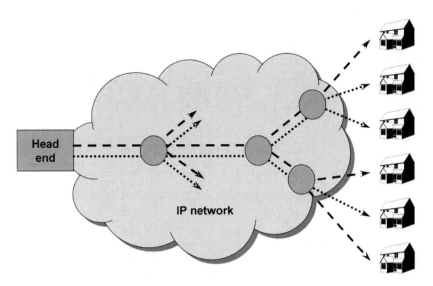

**Figure L-3.** Multicasting for broadcast TV.

deliver one to every home watching that particular television channel. The use of multicasting minimizes the bandwidth required at the head end, because it is only necessary to transmit one copy of each television channel. Figure L-3 illustrates the principle for two separate television channels (one channel is shown dotted, the other is dashed).

If the viewer wishes to change channel, then their set-top box must disconnect from one multicast group and connect to another. The sequence of events is illustrated in Figure L-4.

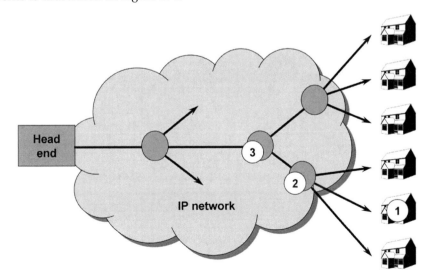

**Figure L-4.** Changing channel in a multicast network.

1. The viewer presses a button on their remote control to select a different channel. The set-top box generates an IP packet containing a request to join the new multicast group.
2. If the required multicast group is not present at the first router that the packet encounters (because none of the homes connected to that router is currently watching the channel), then the request message is sent on its way upstream.
3. The message eventually reaches a router where the multicast group exists. An additional branch is added to the multicast, thereby causing video packets to flow downstream to the viewer that selected them.

This can be a rather slow process. In order to keep channel changing times short and consistent, each television channel is sometimes carried all the way to the edge of the network (Node 2 in Figure L-4) even if nobody connected to that node is actually watching it.

IPTV appears to have a bright future. However, broadband services based on copper telephone lines are not the ideal way to deliver IPTV because television viewers have expectations that copper cables struggle to meet. For example, a typical telephone line is considerably less reliable than a main television transmitter, so viewing is likely to be disrupted more frequently. Furthermore, as television migrates from standard definition to high definition, copper cables may finally hit their bandwidth limit. In many parts of the world, we are starting to see television services being delivered to the home on optical fiber rather than copper cable.

## Appendix M  GSM networks

So, how does a GSM mobile network work? Some of the key components are illustrated in Figure M-1.

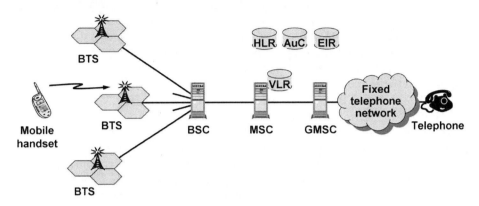

**Figure M-1.**  GSM network.

Each GSM handset contains a Subscriber Identity Module (SIM card), which uniquely identifies the subscriber to the network. The SIM card contains a digital chip that stores information such as the mobile telephone number, details of service subscriptions, preferences, text messages, address book entries and a PIN number. And there is no need to stop there! In Finland, the SIM card is used to store a Citizen Certificate—a government-guaranteed electronic identity that can be used to gain access to a range of digital services. Since all of this information is stored on the SIM card and not in the handset itself, a subscriber can transfer their SIM card from one mobile phone to another without losing any stored information. Conversely, the subscriber can change network operators without buying a new handset by simply changing the SIM card.[1]

When you turn your mobile on, it establishes communication with a nearby Base Transceiver Station (BTS), as shown in Figure M-1. A BTS is normally relatively "dumb"—it relies on a Base Station Controller (BSC) to provide the necessary intelligence. The BSC is responsible for managing issues such as transmit power levels, handoffs between cells and the allocation of radio channels to mobiles. It also acts as a traffic concentrator, multiplexing the large number (sometimes hundreds) of low-speed links coming from BTSs under its control into the small number of higher-speed transmission links that connect back to a Mobile Switching Center (MSC).

The MSC provides call routing but it is far more than a telephone exchange—it also plays a vital role in managing issues such as mobility, cell handoffs,[2] fraud prevention and roaming between GSM networks. Since each cell in the network is linked back to a specific MSC, each MSC is effectively responsible for the operation of the mobile network in a defined geographical area. An MSC that provides an interface between the mobile network and a fixed network is known as a Gateway MSC, or GMSC.

Each MSC is linked to the Home Location Register (HLR), which is the central database where information relating to the network's subscribers is stored. Every active SIM card issued by the mobile network operator has its own HLR entry. This entry holds information such as the SIM's unique identifier, the corresponding mobile telephone number(s), the services that the subscriber is permitted to use, configuration options and information relating to the subscriber's current location. This is pretty important information; if the HLR dies, the network dies with it.

Each MSC also has access to a Visitor Location Register (VLR). While there is only one HLR for the whole network, each MSC normally has its own private VLR. If a subscriber roams into the area controlled by an MSC, the HLR information relating to that subscriber is loaded into the corresponding VLR. By providing temporary storage for this information, the VLR prevents the HLR from becoming a network bottleneck. The VLR also stores information that keeps track of where the subscriber is within the area controlled by the MSC—even if no call is in progress.

There are two other important databases used by the MSC: the Equipment Identity Register (EIR) and the Authentication Centre (AuC). The EIR holds a unique identifier for each mobile handset that is banned from the network, thereby enabling the network operator to suspend service to stolen handsets. The AuC holds a copy of the secret key stored in each SIM card and this enables the MSC to determine whether a SIM Card is genuine.[3] In effect, the EIR is used to check the mobile handset, while the AuC is used to check the SIM card.

A GSM network as described would be pretty good at carrying voice but not much good at carrying data. Figure M-2 illustrates a GSM network with enhanced data capabilities provided by GPRS.[4]

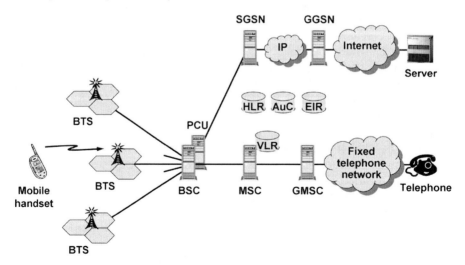

**Figure M-2.**  GSM network with GPRS.

If you compare Figure M-2 with Figure M-1, you will see that three new components have appeared: the PCU, the SGSN and the GGSN. These new components carry data traffic between the mobile handset and an IP network such as the Internet. Voice traffic is unaffected by this change and continues to be handled by the BSC, MSC and GMSC.

The Packet Control Unit (PCU) is generally implemented within the BSC and it manages the data traffic coming from mobile handsets in the area served by the BSC. One of the PCU's primary roles is to separate data traffic from voice, with the data being sent off to a nearby Serving GPRS Support Node (SGSN). The SGSN has a similar role in the packet switched part of the network to that of the MSC and VLR in the circuit switched part of the network; it keeps track of where a handset is located as the user moves around its area and it will perform a handoff to a neighbouring SGSN if that becomes necessary. Data heading for a destination outside the mobile network is sent to the Gateway GPRS Support Node (GGSN),[5] which acts as an interface to the Internet or to some other IP network. Minor modifications are required to the HLR, AuC and EIR to support GPRS, but these can generally be implemented by downloading a software upgrade.

## Appendix N   Wideband CDMA

GSM networks use a mixture of Time Division Multiplexing and Frequency Division Multiplexing to keep users apart on the radio link between the mobile phone and the BTS.[1] The available radio bands are divided up into a number of separate carrier frequencies[2] and each carrier frequency provides digital capacity for eight separate voice channels.[3] Within a particular network cell, some GSM users will be operating on different carrier frequencies (Frequency Division Multiplexing), while others will be sharing the same carrier frequency but transmitting at different moments in time (Time Division Multiplexing). To prevent interference, GSM users in the same cell can never be allowed to transmit on the same frequency at the same time. However, surprising as it may seem, this rule does not apply in 3rd Generation mobile networks. Many 3G networks use a technique called Wideband CDMA (W-CDMA),[4] which allows the same set of carrier frequencies to be used in every cell in the network. This obviously simplifies the frequency planning required, but why doesn't it lead to massive interference?

The short answer is that interference does occur, but it occurs in a way that allows the transmitted data to be recovered at the receiver. Explaining this apparently impossible feat normally requires some rather scary mathematics, but we are going to use a much simpler approach to illustrate some basic principles. Let's consider a cell serving just four users (Andy, Bob, Cathy and Diana) and let's assume that each of them wants to transmit 1 bit of data at a particular moment in time. In order to keep these four transmissions apart, each user is allocated a separate "chip code". Although the codes are made up of binary digits in the sense that each digit can have two possible values, it turns out to be more convenient to use 1 and –1 instead of 1 and 0 to represent the two possible states. This is illustrated in Figure N-1.

| TRANSMITTER | User | Transmit data | Code for 0 | Code for 1 | Transmit signal |
|---|---|---|---|---|---|
| | Andy | 1 | -1 -1 -1 -1 | 1 1 1 1 | 1 1 1 1 |
| | Bob | 1 | -1 -1 1 1 | 1 1 -1 -1 | 1 1 -1 -1 |
| | Cathy | 0 | -1 1 1 -1 | 1 -1 -1 1 | -1 1 1 -1 |
| | Diana | 0 | -1 1 -1 1 | 1 -1 1 -1 | -1 1 -1 1 |

Received signal
= sum of individual   0  4  0  0
transmit signals

| RECEIVER | User | Code for 1 | Received signal | Code x signal | Add results | Received data |
|---|---|---|---|---|---|---|
| | Andy | 1 1 1 1 | 0 4 0 0 | 0 4 0 0 | 4 | 1 |
| | Bob | 1 1 -1 -1 | 0 4 0 0 | 0 4 0 0 | 4 | 1 |
| | Cathy | 1 -1 -1 1 | 0 4 0 0 | 0 -4 0 0 | -4 | 0 |
| | Diana | 1 -1 1 -1 | 0 4 0 0 | 0 -4 0 0 | -4 | 0 |

**Figure N-1.**  Decoding a CDMA signal.

If Andy wishes to transmit a "1", his mobile phone will actually transmit Andy's code (1, 1, 1, 1). If Andy subsequently wishes to transmit a "0", his mobile phone will actually transmit an inverted form of his code (–1, –1, –1, –1). Notice that these codes are unique to Andy; if Diana wishes to transmit a "1", her mobile phone will transmit her code (1, –1, 1, –1).

The last column of the transmitter table in Figure N-1 shows the code transmitted by each of the four mobiles in the cell. However, they are all transmitting at the same time on the same frequency, so interference will occur. If the four signals happen to be correctly aligned, then the signal received at the base station can be calculated by simply adding together the contributions from each of the four mobiles. In this particular example, the signals sum to zero (i.e. they cancel each other out) in three of the four time intervals, giving the result: 0, 4, 0, 0. In spite of this cancellation, we can still recover the data that was transmitted by the four users so long as we know the codes that they were using. We do this by multiplying the received signal by the user's code and then adding the results. If the result is negative, then the piece of data transmitted by the user was a "0"; otherwise, it was a "1". This process of multiplying two sets of data and then adding the results is known as "correlation". Figure N-2 illustrates how correlation is used to recover Cathy's data.

| | |
|---|---|
| Received signal | 0 4 0 0 |
| x Cathy's code | 1 -1 -1 1 |
| = result. | 0 -4 0 0 |
| Add digits to recover | |
| transmitted data | -4 |
| (0 if –ve; 1 if +ve)  → | 0 |

Figure N-2. Correlation.

Correlation is often used in signal processing to detect the presence of a known signal that is hidden in large amounts of noise—and that, if you think about it, is what is going on in W-CDMA. If you wish to receive Bob's signal, then the signals from Andy, Cathy and Diana can be regarded as a form of background noise. It is rather like having a conversation at a crowded cocktail party; there may be many separate conversations going on around you, but you concentrate on your own conversation and all the other chatter merges into the background noise.

We have now demonstrated that it is possible for multiple users to transmit simultaneously on the same frequency without data being lost. However, real-life realizations of W-CDMA are considerably more complex than the example given above because they have to overcome a number of practical difficulties. For example, it was assumed in the example that the four mobiles were synchronized so that their transmissions could simply be added together, but this synchronization is hard to achieve when mobiles are moving around. Why, then, does anyone bother with W-CDMA when alternatives such as TDM and FDM seem to be so much simpler? Some important reasons are as follows:

- Data traffic tends to be bursty. W-CDMA can handle bursty traffic more efficiently than TDM or FDM.
- W-CDMA makes more efficient use of radio spectrum.
- Network planning is much simpler, because W-CDMA allows all the base stations in an area to operate on the same radio frequencies.

As a result of these advantages, W-CDMA is widely used in 3G mobile networks.[5]

## Appendix O  Network reliability

Table O lists a number of different network problems that can occur. An analysis of network reliability needs to consider all of these possible types of failure.

**Table O.**  Network problems.

| Problem | Possible causes |
| --- | --- |
| Hardware failures and accidental damage | Electronic equipment failures or damage to cables caused by vibration, electromagnetic interference, lightning, power supply problems, ice, wind, floods, fires, earthquakes or rodents. Satellites can be damaged by space debris or solar flares |
| Deliberate sabotage | Caused by bombs, computer viruses, cable cuts or arson |
| Software failures | Caused by software bugs or problems that arise during software upgrades |
| Operational errors | Caused by network configuration errors, network planning errors, process problems or errors in records |

As an example, failures in electronic equipment typically follow a "bathtub curve", as shown in Figure O.

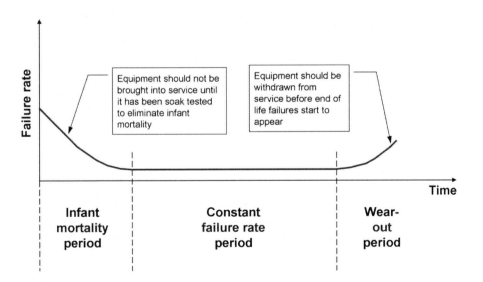

Figure O.  Bathtub curve.

Immediately after manufacture, there is a period of "infant mortality" during which component faults, manufacturing defects and other weaknesses come to light. This is then followed by the constant failure rate period, during which there is a low level of random failures. Eventually, the failure rate starts to rise again as the equipment reaches the end of its working life. Clearly, equipment should not be brought into service until it has been soak tested to eliminate infant mortality and should be withdrawn from service before end-of-life failures start to appear.

Although all networks experience occasional equipment failures, many problems can be prevented by storing the equipment in high-quality data centers where it is protected from vibration, temperature changes, humidity, power supply glitches, fire, flooding and other hazards. Data centers provide a highly secure environment in which the equipment is protected from physical attack, sabotage and all forms of unnecessary human intervention. As a general rule, electronic equipment prefers a comfortable but really boring existence. Even turning it on and off can stress it and shorten its life!

Data center facilities for storing equipment can be provided cost-effectively for core network nodes because the number of separate sites required is usually quite small and there is a significant amount of equipment at each site. However, it is often impractical to provide these facilities in the access network. Some access equipment may have the misfortune to be located in a street cabinet, where it will be subjected to temperature extremes, humidity and occasional vandalism.

Protecting cables from damage is usually more difficult than protecting equipment. Buried cables are regularly cut by utility companies digging up the street, while submarine cables are dragged up by trawlers. Overhead cables are vulnerable to hazards ranging from falling trees to high-sided vehicles. Fiber optic cables carried on high-voltage pylon lines have been damaged by shotgun pellets, low-flying aircraft and even—amazingly—by vandalism. Although network operators take precautions to minimize cable cuts, eliminating them completely is not a realistic possibility. For this reason, the time taken to repair a cable after it has been damaged is a critical factor in the calculation of service availability.

Modern networks are becoming increasingly software-based. Even if the network hardware is beautifully engineered, the whole thing can fall apart if the software contains bugs. In the many millions of lines of code that control modern networks, it is almost inevitable that a few software bugs will be lurking somewhere and these bugs can occasionally cause network outages that disrupt service to millions of customers.[1] A classic example of this occurred on January 15, 1990, when large parts of AT&T's North American long-distance network crashed. A few months previously, the software in 114 tandem switching centers had been upgraded. Unfortunately, there was a single defective line of code in one of the failure recovery routines.[2] The crash lasted for 9 hours and 70 million phone calls were lost. Some estimates put the eventual cost of the disaster to AT&T as high as $1 billion. That is a *very* expensive software error!

Although this was one of the more widely publicized software failures, AT&T

is by no means the only network operator to have experienced a catastrophic failure caused by software problems. For this reason, network operators tend to exercise extreme caution before introducing any software upgrades into their networks.

A number of network failures that are blamed on "software glitches" are actually caused by operator errors. Software-based networks require a considerable amount of configuration data to be entered before they will operate properly. If any of this data is entered incorrectly, the results can be disastrous and stories about a seemingly innocent configuration change causing a major network to collapse in a heap are all too common in the industry. Given that human error is a fact of life, what can be done to minimize this problem?

In many cases, operational errors are attributable to lack of training, poor supervision, badly defined processes or inadequate record keeping. Mistakes can also occur because a temporary "work around" repair has been made to fix an earlier problem and the staff in the network management centre have not been made aware of what has been done. When it comes down to it, most operational errors are ultimately attributable to inadequate investment or lack of effective management. The operational management team have a key role to play in preventing bad working practices and all cases of operator error need to be properly investigated so that lessons can be learned.

It is clearly important that operations staff are not asked to carry out tasks that are beyond their capabilities. One way to prevent this is to give each operator a defined set of capabilities on each computer system that they use. This ensures that the system will only allow them to make configuration changes that are considered to be within their capabilities. A junior operator might be authorized to make changes that could affect a small number of customers, but only the most experienced operators would be allowed to make changes that could potentially affect thousands of customers. The system might also restrict changes to times when the network is not being heavily used, thereby minimizing the impact of any mistakes.

## Appendix P   Availability

There are a number of different ways of measuring the quality of a network, but service availability is one of the most important. Availability measures the proportion of time that a service is working properly, so a service with a 99.999% availability target[1] can be defective for up to 5.3 minutes per year.[2] The concept of availability is simple enough, but the devil's in the detail. For example, does a voice channel count as available or unavailable if there is an annoying level of background noise? Is a data channel available if it is introducing occasional bit errors? Even if we can agree whether a service is working or not, the way in which network faults occur can also be significant, as illustrated in Figure P.

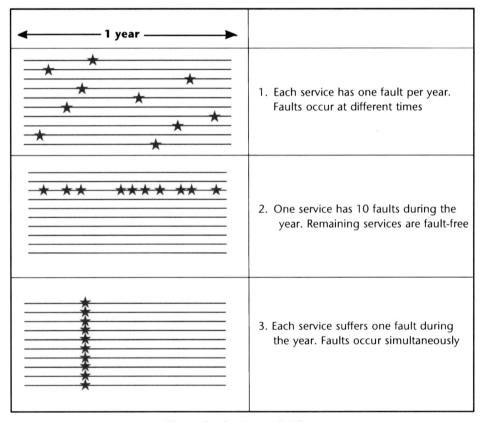

**Figure P.**   Service availability.

The three parts of Figure P illustrate the annual performance of 10 telephone lines serving a bank branch. In the first case, the faults are randomly distributed and there are always at least nine telephone lines available for use. In the second case, the bank experiences an appalling level of service on one telephone line, while the other nine lines are fault-free for the whole year. In the third case, all

10 faults occur simultaneously, so there is a period of time during which the bank is unable to communicate with its customers. All three situations show the same number of faults during the course of a year, but the customer experience is very different.

Now let's consider the same three situations from the point of view of the network operator. In the first case, the faults are evenly distributed over the course of a year, so the same technician can be used to repair all 10 faults. Furthermore, calls from customers wishing to report a fault are evenly distributed, so the need for call-center staff is minimized. Strictly speaking, the same advantages also apply in the second case, although the dreadful level of service on one particular line might persuade the customer to switch to another network. In the third case, all the faults occur simultaneously, so the repair technicians and the call-center staff could easily be overwhelmed. From both a customer and a network operator point of view, this is a situation that must be avoided at all costs.

Unfortunately, some core network faults have exactly this type of characteristic. An optical fiber in the core network may be carrying thousands of simultaneous telephone calls, so a single cable cut can cause widespread disruption. For this reason, a high proportion of investment in network resilience is spent to protect the core network, whilst little or nothing is spent in the access network. The telephone line connecting a domestic customer to their local exchange typically has no resilience at all. It's not that the telephone company doesn't care about its domestic customers—it's simply a case of spending the money where it will do the most good. Investing in core network resilience is a cost-effective way of improving service for all the customers on the network, while providing resilience on domestic telephone lines would be extremely expensive and would lead to intolerably high phone bills.[3]

## Appendix Q   Error detection and correction

For various reasons, packets occasionally become corrupted as they travel across a network and error detection techniques are used to spot when this occurs. Parity checks provide a simple example of how this can be done. Each byte of data carries an additional bit called a parity bit and the rule is that the number of ones in each byte—plus the parity bit—must be odd.[1] A couple of examples illustrate the principle (Table Q-1).

**Table Q-1.**  Parity check examples.

| Data | Parity |
|------|--------|
| 10011000 | 0 |
| 00101110 | 1 |

If there is a single bit error anywhere in the data or the parity bit, then there will be an even number of ones and the receiver will be able to detect that the data has become corrupted. If, however, there are two bit errors in the data and parity bit, then there will once again be an odd number of ones and the parity check will fail to indicate an error. A second parity bit could be added to protect against the possibility of two errors, but it might still fail to detect the occurrence of an even larger number of errors. Parity bits represent a form of redundancy— they are there to protect the data rather than to carry useful information—so the level of the parity overhead must be chosen to suit the value of the data. As always, the aim is to minimize the cost of this overhead while providing an acceptable level of protection.

To detect errors in a block of data, a checksum can be used. A very simple method of calculating a checksum would be to add each byte of data in the block, as illustrated in Table Q-2.

**Table Q-2.**  A very simple method of calculating a checksum.

|  | Binary | Decimal |
|------|--------|---------|
| Data | 00001000 | 8 |
|  | 00011000 | 24 |
|  | 00000011 | 3 |
|  | 00001010 | 10 |
|  | 00010101 | 21 |
| Checksum | 01000010 | 66 |

The checksum is calculated at the transmitter and transmitted along with the block of data. At the receiver, the checksum is recalculated and compared with the transmitted version. If the two checksums are not the same, then errors have occurred during transmission. Unfortunately, however, this simple form of checksum cannot detect whether the order of the bytes has been changed or whether zero-valued bytes have been added. More sophisticated types of checksum can be used to address such issues but the calculations are obviously more complicated.

On the Internet, a checksum is used to detect errors in a packet header. The value of the checksum is calculated at the transmitter and is stored in the packet. The value of the checksum is then recalculated at each node that the packet passes through. The packet will be discarded at any point if the calculated checksum does not match the checksum stored in the packet and the transmitter will subsequently re-transmit the packet.[2] This technique of automatically re-transmitting corrupted messages has been used for many years to provide error-free data communication on noisy radio links. Indeed, the idea of requesting re-transmission of garbled messages has been around since the earliest days of the telegraph.

When communicating with spacecraft in deep space (such as NASA's Voyager and Mariner series), the received signal is extremely weak and errors are to be expected. However, the transmission delay from Earth to the probes can be as much as half a day, so requesting re-transmission of a corrupted message would be a very cumbersome process. Transmission delay is also an issue for satellite television channels because re-transmitted data would probably not arrive until after the need for it has passed. In these situations, error correction techniques are needed. Error correction has many similarities to error detection but it doesn't just detect errors—it fixes them as well! As you might expect, error correction requires additional (redundant) bits to be transmitted and the number of errors that can be corrected depends upon the number of additional bits that are transmitted.

To the layman, the idea of a coding technique that can correct its own errors seems almost magical, but the following simple example illustrates one way in which it can be done. Let's suppose that you want to transmit the binary value 11010. Instead of transmitting it just once, you send it three times in succession:

```
11010
11010
11010.
```

Now let's assume that a single bit error occurs during transmission, so the received data becomes:

```
11010
11010
01010.
```

Since the three values are not the same, we know that at least one error has occurred. Closer inspection reveals that the first two values are the same, while the third one is different. Since the probability of a single error is higher than the probability of two errors, we conclude that the error must lie in the third value.

Comparing the third value with the other two values reveals that the error has occurred in the first bit (underlined).

This code is capable of correcting multiple errors. To illustrate this, let us suppose that the received data looked like this:

    11011
    11010
    00010

There are now three errors hidden in the data. If you compare the corresponding bits in each value and assume that the majority are correct, you should be able to find the errors and correct them. Notice that correcting a bit error in binary data simply requires the bit to be inverted: a 1 becomes a 0 and a 0 becomes a 1.

This particular code is pretty inefficient because it increases the bit rate of the original data by a factor of three. It can always correct a single error in the data and can sometimes (but not always) correct multiple errors.[3] Far more efficient error correction techniques have been devised but they require some sophisticated mathematics and they are highly optimized for the applications in which they are used. For most telecommunications applications, error detection combined with re-transmission of corrupted packets is a much better solution than error correction.

# Notes

## Introduction

1. A beacon was lit when the Armada was first sighted. This caused the next beacon in the chain to be lit and the message was passed from beacon to beacon along the coast. At Beachy Head, the line of beacons turned inland to bring the message to London.
2. The word "telecommunication" comes from the Greek word "tele" (far away, at a distance) and the Latin word "communicare" (to impart, share).

## Chapter 1: The birth of an industry

1. One proposed amendment had sought to link the experimental telegraph with experiments in hypnosis.
2. Morse had suffered personally from the painfully slow communications of the time. In February 1825, he had received a letter in Washington telling him that his wife, Lucretia, was gravely ill. However, the letter had taken 4 days to reach him. By the time he returned to his home in New Haven, Connecticut, his wife was already dead and buried.
3. An electromagnet is simply a coil of wire wrapped around a core made of some suitable magnetic material, such as iron. Electromagnets have the very useful property that they are only magnetic when an electric current is flowing through the coil.
4. The "International Morse Code" shown in Table 1 actually differs in some respects from the original developed by Morse. The code sequences for 11 characters were changed in 1848 and further changes occurred in 1865. The extraordinary longevity of Morse Code is illustrated by the fact that the letter "@" was added to the code in 2004 to enable email addresses to be transmitted.
5. Numbers, Chapter 23, Verse 23.
6. A Leyden jar was a device that was used to store electrical energy. In the terminology of modern electronics, it would be referred to as a capacitor.
7. It appears that the first use of the word "telegram" in England occurred in 1852. It was opposed by some pedants, who claimed that the correct derivation from the Greek should have been "telegrapheme".
8. Some relays simply left an indentation on the paper. Others used paper that had been chemically treated and marks were made by passing an electric current through the paper.

9. *Harpers New Monthly Magazine*, 1873.
10. In 1951, only 1.5 million UK households were on the telephone. More than half of all UK homes still did not have a telephone by the end of the 1960s.
11. Operators of telephone networks used to refer to their customers as "subscribers". In the United Kingdom, the practice was discontinued in 1959 but the term has taken a long time to die. It appears in this book because many of the developments that we will discuss happened while the term was still in widespread use.
12. Cooke and Wheatstone were subsequently allowed by the railway company to extend their system as far as Slough on condition that railway messages were carried free of charge.

## Chapter 2: The telegraph goes global

1. Morse demonstrated that a 2,000-mile telegraph line could be made to work by connecting together 10 separate telegraph lines running between London and Manchester. However, as Field was to discover, the electrical properties of submarine cables are very different from those of overhead telegraph wires.
2. In Baltimore, a 5 × 2.5-feet copper ground plate was immersed in the sea at Pratt Street docks. In Washington, a similar plate was buried in the cellar of the Capitol building.
3. Edison's quadruplex telegraph managed to transmit two simultaneous messages in each direction by using the strength of the signal to carry one message while the polarity of the signal carried the second message. However, this approach depended upon finding different characteristics of the electric current to carry each additional message, and it was not clear that this system could be enhanced very much further.
4. This apparatus was based on an artist's canvas stretcher. Morse had started his career as a portrait painter.
5. As in the case of the House printing telegraph, the Stock Ticker used a rotating disk to produce readable text. Its name derived from the ticking sound made as the disk clicked round to select the next letter.
6. Teleprinters are often referred to generically as "teletypes", rather in the way that vacuum cleaners are referred to as "Hoovers". The Teletype Corporation of Skokie, Illinois, produced its first general-purpose teletype in 1922.
7. During the second half of the twentieth century, the teleprinter acquired a new lease of life as an interactive computer terminal for mainframes and minicomputers. Commands were entered in response to a prompt character, giving rise to the command line interface that subsequently appeared in operating systems such as Microsoft's MS-DOS.
8. Bain was very familiar with electromagnets and pendulums. His first patent, in 1841, was for an electric clock in which electromagnets were used instead of springs or weights to keep the pendulum swinging.
9. In Bain's approach, the flow of current had indicated black.

## Chapter 3: A gatecrasher spoils the party

1. At a time when his experiments were not going well, Bell had visited Joseph Henry in Washington to seek advice. Professor Henry was, by then, aged 78 and was one of the leading figures in American science. Bell was 28 and virtually unknown. However, Henry was as generous with his help to Bell over the invention of the telephone as he had been 35 years earlier to Samuel Morse over the invention of the telegraph. Indeed, many claimed that Henry—rather than Morse—had actually been the true inventor of the telegraph, since he had demonstrated a system operating over a mile of cable as early as 1831. However, Henry believed that such discoveries should be made freely available for the public good rather than being locked away for personal profit by patents. The unit of electromagnetic induction was later named the Henry in his honour.

2. In 1956, Lerner and Loewe converted Shaw's play into the musical *My Fair Lady*.

3. A. G. Bell, *The Multiple Telegraph*, 1876.

4. In his patent, Bell uses the term "armature" instead of reed. To relay engineers, there is a difference between a reed and an armature, but the rest of us can safely consider them to be the same thing.

5. The same is true of the current in a standard telegraph circuit, in which the Morse key is either open or closed.

6. Bell referred to "undulatory" and "pulsatory" currents in his telephone patent, so the same terminology has been adopted here. In modern terms, pulsatory currents are found in digital systems, while analog systems work with undulatory currents.

7. Fagen, M. D., ed.: *A History of Engineering and Science in the Bell Telephone System: The Early Years (1875–1925)*, Bell Telephone Laboratories, New York, 1975, pp. 7–8.

8. If the reed is constructed using a permanent magnet, then the telephone can be demonstrated without the need for a battery.

9. Since the circuit is also carrying a steady current driven by the battery, the induced current driven by the vibrating reed will actually appear as a ripple superimposed upon the steady current. This does not change the basic principles of the design.

10. Bell had chosen his tuned reeds so that they would each respond strongly to one particular frequency (its resonant frequency) and reject other frequencies. However, a resonator can be persuaded to respond to a broader range of frequencies if damping is applied. In this case, the pressure of Bell's ear against the reed was supplying the necessary damping.

11. There is nothing fundamentally wrong with the design of the gallows telephone, and it can be made to work. However, it appears that the resistance of the electromagnets was too low for efficient operation.

12. It is generally accepted that these were the first words ever to be spoken and received over a telephone. Watson later claimed that they were spoken in

an urgent tone of voice because Bell had just upset battery acid over his clothes.

13. Alexander Graham Bell family papers, Library of Congress.

14. The detector worked, but it failed to find the bullet. It was later discovered that the detector had been confused by the metal bed springs under the President's mattress.

15. Bell mistakenly believed that sheep with extra nipples would give birth to more lambs.

16. A few years later, Bell was able to reject an offer of $25 million for the patent from the same company.

17. A caveat filed with the Patent Office indicated an intention to submit a patent application as soon as the invention had been more fully developed.

18. In October 1946, Sir Frank Gill, the Chairman of STC, asked whether STC could examine several original Bell and Reis instruments held by the Science Museum in London as part of plans to celebrate the centenary of Bell's birth. To everyone's surprise, it was found that Reis's transmitter could transmit intelligible speech. STC insisted that these results were kept secret because they did not wish to jeopardise their good relationship with AT&T (the commercial arm of the Bell system) by suggesting that Alexander Graham Bell did not invent the telephone. It appears that the Reis telephone works as a loose-contact microphone, but it only does so when the sound is so soft that the contact is never actually broken. This suggests that the Reis device may indeed have carried intelligible speech 16 years before Bell's patent, but it did so entirely by accident and in a way that its inventor did not understand. The fact that the invention was based upon an accidental discovery would not be a problem—Bell's invention of the telephone was inspired by the plucked reed incident—but Reis (unlike Bell) never appreciated the significance of his discovery.

## Chapter 4: Early telephone networks

1. At the time the advertisement was written, the only telephone line in the world ran between Charles Williams' workshop in Court Street, Boston (where all Bell's early telephone apparatus had been made) and Mr Williams' home in Somerville, Massachusetts.

2. For North American readers, a local exchange is a "central office".

3. To some older readers, this may sound like a party line. This was a telephone line shared between two subscribers. However, the idea of a party line was not to enable the two subscribers to talk to each other, but to avoid the cost of installing an additional telephone line to the local exchange. Since the line was shared, there would be times when one subscriber was unable to make telephone calls because the line was in use by the other subscriber. Eavesdropping was also a problem.

4. It is sometimes claimed that the first switchboard was set up in Boston in

1877. A young man named E. T. Holmes, who ran a small burglar alarm business, became the first person to offer a telephone service. He used six telephones in his office, which could be connected as required to six burglar alarm wires. Any two of the six wires could be linked at the office to establish a telephone connection.

5. The telephone companies tended to use girls as telephone operators because boys were considered to be too boisterous and rude.

6. The concept of automatic switching had originally been patented in 1879 by M. D. Connolly, T. A. Connolly and T. J. McTighe, and a system supporting eight subscribers was exhibited at the Paris Exposition of 1881. Although they introduced a number of important concepts that would be used in later designs, their system was crude and impractical. A number of other inventors were active in this field at about the same time, but Strowger was the first to come up with a practical implementation of automatic switching that could support a commercial service.

7. As the telephone system grew, switchboard operators became very busy at peak periods and delays in answering calls became an increasing source of annoyance. Calls were often routed to the wrong destination and conversations were occasionally cut off by mistake. When his system made its debut, Strowger claimed that his telephone exchanges were "girl-less, cuss-less, out-of-order-less and wait-less".

8. Although we talk about seven "pulses" being sent when the number 7 is dialled, it would be more correct to say that the line current is interrupted seven times. Line current is also interrupted when the telephone is not being used, because there is no point in draining the telephone exchange batteries unnecessarily. This means that it is often possible to dial a number on a pulse dialing telephone by rapidly operating the switch that detects when the receiver has been replaced. For example, to dial the number 7, the switch would be pressed seven times in quick succession. In the United Kingdom, it used to be possible to make free calls from some public telephone boxes using this technique, although people caught doing this were likely to be charged with "stealing electricity".

9. Things moved more slowly in the United Kingdom than in the United States. It was not until 1912 that the British Post Office finally deployed three automatic telephone exchanges that could handle local calls automatically.

10. By 1966, STD covered about two-thirds of the United Kingdom.

11. The term "phreak" is apparently derived from a combination of the words "freak", "phone" and "free".

12. The connection between the two long-distance exchanges in Row C is shown dotted because there may be additional exchanges in between. In the early days of the telephone network, long-distance connections were so expensive that it made sense to use them as efficiently as possible. This led to the development of a large number of small telephone exchanges distributed widely around the country and a multi-level exchange hierarchy became

necessary to manage this complicated arrangement. As modern technologies such as fiber optics dramatically reduced the cost of long-distance network connections, networks evolved towards fewer but bigger exchanges arranged in a much flatter hierarchy. In the modern BT network, all the long-distance exchanges are fully interconnected, so a long-distance national call should never need to pass though more than two long-distance exchanges.

13. The Triode was the forerunner of the vacuum tubes that were used in television sets until relatively recent times. Although vacuum tubes were gradually displaced by the transistor, they are still used in certain specialist applications. For example, many rock guitarists insist that the distortion produced by an overdriven valve amplifier sounds better than anything that semiconductor technology can produce.

14. In fact, even the telegraph struggled on routes as long as this. The trans-Atlantic cable was actually built in three sections: London to Valentia Island (Ireland), Valentia Island to Trinity Bay (Newfoundland) and Trinity Bay to New York. After a signal had crossed the 1,896-mile central section of the cable, it was too weak to be regenerated using a relay. Highly sensitive mirror galvanometers were used to read the incoming message and the message was then re-keyed by a telegraph operator.

15. About 50 amplifiers were required at regular intervals along the cable to span the Atlantic. Since they had to operate at great depths, extremely high reliability was required.

16. Many hi-fi enthusiasts regard the analog background noise on a vinyl LP as preferable to the sterile perfection of a digital CD.

17. Casson, Herbert N.: *The History of the Telephone*, A. C. McClurg & Co., Chicago, 1910, Chapter 4.

18. The metallic or "two-wire" circuit was patented by Alexander Graham Bell in 1881. However, the conversion of overhead lines from one wire to two was both complex and expensive, and was not completed until 1900.

## Chapter 5: Going digital

1. To this day, radio amateurs use Morse Code to communicate over low-quality radio channels that cannot carry intelligible speech.

2. During the early years of the Second World War, an analog scrambling system was used on the radio-based telephone "hotline" between Prime Minister Winston Churchill and President Franklin Roosevelt. Although this system was sufficiently secure to protect against casual eavesdroppers, it was not sufficient to defend against determined German code-breaking and unscrambled transcripts of the wartime leaders' conversations would be on Hitler's desk almost as soon as the conversation had finished. However, this analog system was replaced in 1943 by a system employing digital encryption and the digital code resisted all German attempts to break it.

3. This term derives from the fact that the waveform has changed its appearance and so it can be considered to have adopted an alias.
4. According to the Nyquist Criterion, it should be possible to pass frequencies that are above 3.4 KHz so long as they are below 4 KHz. In practice, however, complex filters are required to do this and they are expensive to implement.
5. The process of converting an analog signal to digital is usually referred to as "A to D conversion". The converse process is called "D to A conversion".
6. Many years after Reeves' PCM patents had been granted, a little known patent by Paul M. Rainey dating from 1926 was found to contain many of the key features of PCM. This naturally led to claims that Reeves was not the inventor of PCM, although it is generally accepted that he was the first to build a working PCM multiplexer. Rainey's patent describes a method of transmitting facsimile over a telegraph network using a form of PCM. The patent does describe key components of the PCM system such as quantization, but it does not recognize the need for filtering prior to sampling and does not suggest that there are any applications for the technique beyond the transmission of facsimile images. In contrast to this, it is clear from Reeves' patent that he appreciated the full technical requirements of PCM and the wider implications of what he was doing. It is significant that Bell System engineers had forgotten Rainey's work by the time that Reeves filed his patent.
7. After 145 years, Western Union finally stopped carrying telegrams on January 27, 2006.
8. In the 1930s, the development of teleprinters that were able to operate across telephone channels meant that there was no longer any need to maintain a physically separate telegraph network. The remaining telegraph lines were converted into telephone trunks to provide the telephone network with additional capacity.

## Chapter 6: A bit of wet string

1. If this is not done, the signal will be lost in the noise that is inevitably present in an analog communication channel. Noise may not be directly visible in a digital communications channel but, as Shannon demonstrated, it limits the bit rate that the channel can support.
2. The techniques used to eliminate echoes on long telephone lines are discussed in Appendix D.
3. In 1879, Professor David Hughes noticed that a clicking noise occurred in his home-made telephone whenever he used his induction balance. Hughes found that the induction balance had a loose contact and that the clicking went away when the contact was fixed. He correctly deduced that radio waves were emanating from sparks at the loose contact. Hughes devised a clockwork device to generate the sparks at regular intervals, thereby producing a regular clicking noise in the telephone handset. When he found that these clicks

could be heard all round his house, he set off down Great Portland Street in London with the telephone held to his ear. (This has been claimed by some to be the first ever use of a mobile phone!) As he later recalled, "the sounds seemed to slightly increase for a distance of 60 yards, then gradually diminish, until at 500 yards I could no longer with certainty hear the transmitted signals". The discovery was demonstrated to members of the Royal Society in February 1880, but he was told that it was merely induction. Hughes was so discouraged that he did not even publish the results of his work and the credit for these discoveries went to others. However, David Hughes achieved the recognition that he deserved in many other areas and he appears in Appendix C as one of the inventors of the microphone.

4. Electrical resistance causes telephone signals to weaken as they travel down the wire. However, thicker wires have lower resistance, so wires as thick as 1/6 inch were used on circuits such as the original New York–Chicago line of 1892.

5. Casson, Herbert N.: *The History of the Telephone*, A. C. McClurg & Co, Chicago, 1910, Chapter 4.

6. Maxwell's Demon was a "thought experiment" (i.e. a hypothetical experiment that could never be realized in practice). It was intended to illustrate a problem with the Second Law of Thermodynamics. Suppose that you have a box containing gas at a constant temperature and you introduce a partition to divide the box into two separate chambers. Now suppose that the partition contains an extremely small trap door that is just big enough to allow a molecule of the gas to pass through. This trap door can be opened and closed on frictionless hinges and it is under the control of an intelligent "Demon". The temperature of the gas is determined by the average speed of its molecules, but some of those molecules will be traveling faster than others. If the Demon only opens the trap door when it detects a faster-than-average molecule approaching from the right, or a slower-than-average molecule approaching from the left, then the average speed of the molecules in the left-hand chamber will gradually increase, while the average speed in the right-hand chamber will gradually fall. This creates a difference in temperature that could be used to do useful work (e.g. by driving a heat engine), even though no energy has been expended in creating the temperature difference. A number of interesting discussions on the significance of Maxwell's Demon can be found on the Internet.

7. Einstein was so inspired by Maxwell that he placed a photograph of him on his study wall.

8. The spark gap initially presents a high electrical resistance, allowing the capacitor to charge. When the breakdown voltage of the air gap is reached, the air molecules in the gap become ionized and a plasma of ions and free electrons is created between the electrodes. This means that the air gap now presents a low electrical resistance and the capacitor discharges. Since the capacitor and the induction coil form a resonant circuit, the discharge takes the form of a damped electrical oscillation.

9. Oliver Lodge established the existence of electromagnetic waves indepen-

dently of Hertz and at almost exactly the same time. However, Hertz was the first to publish his results. As we shall see, Lodge is one of many unsung heroes in the history of radio.

10. Marconi, Guglielmo: "Wireless Telegraphic Communication", Nobel Lecture, December 11, 1909.

11. We previously encountered Lord Kelvin in Chapter 3, when, as Sir William Thomson, he attended the Centennial Exhibition in Philadelphia and was one of the party of dignitaries who marveled at Alexander Graham Bell's newly invented telephone.

12. A transmitter aerial is designed to be good at transmitting radio waves, so any oscillations set up by a spark discharge will quickly die away as the energy is radiated. Ideally, the circuit will also maintain its oscillation between each spark discharge by resonance. However, it was recognized that a circuit can be designed to be a good radiator of energy OR to achieve sustained resonance, but not both. By having two tuned circuits in his apparatus (a highly resonant circuit and an aerial with good radiating characteristics) and then weakly coupling the two circuits, Marconi was able to improve the range and selectivity of his equipment.

13. Marconi, Guglielmo: "Wireless Telegraphic Communication", Nobel Lecture, December 11, 1909.

14. G. Marconi, quoted in *A Science Odyssey: People and Discoveries, www.pbs.org*.

15. Marconi had to abandon his original site at St Johns, Newfoundland, as a result of the hostility of the local telegraph company, which claimed the rights to all forms of telegraphy. In contrast to this, the Canadian government provided active advice and financial assistance for the construction of a high-power station at Glace Bay, Nova Scotia.

16. When the *Titanic* hit the iceberg, another ship—the *Californian*—was only 10 miles away. Its radio operator had gone to bed, having been instructed to shut up by the *Titanic* because he was interfering with their communication with Newfoundland. The First Officer on the *Californian* noticed white flares being launched from the *Titanic*, but thought that they were part of a celebration. To make sure, he tried to signal the *Titanic* with a signal lamp, but received no reply. He did not, however, wake up the radio operator and send a message by radio. If he had done so, it is likely that the death toll on the *Titanic* would have been very much lower. The following year, an international treaty required all ship's radios to be manned 24 hours a day.

17. Pierce is also credited as the inventor of the word "transistor". See *www.pbs.org/transistor/album1/pierce/*.

18. Shannon (see Chapter 5) had demonstrated that the maximum bit rate of a channel was related to its bandwidth. As a rule of thumb, the higher the frequency of the carrier, the higher its potential capacity to carry traffic.

19. This group worked under Alec Reeves, who had invented pulse code modulation in 1937 (see Chapter 5).

20. Kao, K. C. and Hockham, G. A.: "Dielectric-Fibre Surface Waveguides for Optical Frequencies", *Proc. IEE*, **113**(7), 1151–1158 (July, 1966).

21. The target proposed by Kao and Hockham was less than 20 dB per kilometer. The fiber produced by Corning achieved 17 dB per kilometer. Some modern fibers can achieve attenuation below 0.1 dB per kilometer. To illustrate just how pure this is, a block of fiber optic glass 20 kilometers thick would be as transparent as a window pane!
22. *http://nobelprize.org/nobel_prizes/physics/laureates/2009/*.
23. Plastic fibers are used in some applications, but glass is preferred for long-distance telecommunications because of its lower attenuation.
24. Lasers are more powerful than LEDs, but LEDs are cheaper and have longer working lives.
25. Early optical fibers had no external protection and the total internal reflection occurred at the glass–air interface. However, it was found that covering the fiber with a transparent cladding of lower refractive index protected the reflecting surface from contamination and improved performance. The cladding is normally covered with a plastic coating to protect against moisture and physical damage.
26. For a given level of dispersion, there is a trade-off between bit rate and distance. For this reason, fiber transmission systems are often characterized by their *bandwidth–distance product*. For example, a multimode fiber with a bandwidth–distance product of 500 MHz.km can carry a 500 MHz signal for 1 kilometer or a 1,000 MHz signal for 0.5 kilometers.
27. Although single mode fiber has eliminated the principal cause of dispersion, the performance of optical systems is now limited by some rather more exotic forms of dispersion. Chromatic dispersion is caused by the refractive index of the fiber varying slightly for different frequencies of light. Polarization mode dispersion occurs because the propagation velocity of light in the fiber varies slightly, depending upon its polarization. Even with single mode fiber, a point is eventually reached on long-haul systems at which it becomes necessary to convert the optical signal back to an electrical form, regenerate the pulses and convert them back to optical for onward transmission. This is an expensive operation but, fortunately, it is only required on extremely long links.

## Chapter 7: The last mile

1. "Central office" is the American term for a telephone exchange.
2. Quoted in Casson, Herbert N.: *The History of the Telephone*, A. C. McClurg & Co, Chicago, 1910, Chapter 4.
3. In the United Kingdom, BT's Access Network contains more than 120 million kilometers of copper wire—enough to go around the world 3,000 times!
4. ADSL stands for Asymmetric Digital Subscriber Line. The "asymmetric" refers to the fact that the bandwidth provided from the network to the user is much greater than the bandwidth provided in the opposite direction. This improves the performance of the service for web surfing—but not for file sharing and other activities requiring a fast uplink.

5. ADSL uses a form of Frequency Division Multiplexing. The voice channel occupies frequencies from 300 Hz to 3.4 KHz, while the digital data is carried by frequencies above 25 KHz.

6. In rural areas, telephone cables are often installed on telegraph poles (where they are vulnerable to tree branches and weather). In urban environments, they are generally buried underground (were they are vulnerable to people digging up the street).

7. In theory, cable TV access networks could have delivered broadband services using ADSL over their twisted pair telephony cables, but they chose instead to deliver broadband using the coaxial cables that carry their television services. They did this by placing the downstream data (from network to user) in an 8 MHz frequency slot that would otherwise be carrying a television channel. This means that the same data is broadcast to a number of homes in the street. A device called a cable modem is required at the customer site to separate this broadband data from the television signals and to select which packets of data are actually intended for that particular user. Upstream data (from user to network) is placed in a different frequency slot on the cable and each cable modem, in turn, is allocated a short period of time in which to transmit data to the network.

8. The digital bandwidth that can be supported by twisted pair is inversely related to the length of the cable—as some broadband users who live a long way from their local telephone exchange have found to their cost.

9. Of course, high-value business customers can justify the cost of a fiber link all the way to their premises, but domestic customers and small businesses are rather less fortunate. In places like the City of London, network operators often install fiber connections to bandwidth-hungry (and rich) customers such as financial institutions.

10. Ionica's failure is generally attributed to software development delays and poor marketing, suggesting that the company could have been successful if it had started with more mature technology and a better management team. However, there were at least three fundamental problems that would affect any future start-up using a similar business model:

   i.  *High customer acquisition costs.* Microwave radio was used to link each customer to a local base station. Since microwaves require a clear line of sight from the transmitter to the receiver, the customer's aerial had to be mounted high up on the outside of the house. As a result, aerial installation required an expensive site visit by an engineer. Costs were further increased by the equipment required to receive the service and by the cabling and installation costs inside the home. Although the BT network has to maintain a copper cable to each customer, it is likely that the installation costs of this cable were paid for many years ago.

   ii.  *High customer costs.* In a mature market such as the United Kingdom, new operators have to win customers by persuading them to move from existing operators. These customers are unlikely to be loyal; having moved once, they are likely to move again if the service does not live up to their expectations. The high rate at which Ionica's customers defected to other networks meant that de-installation costs were very significant. Furthermore, new operators such as Ionica tend to attract people with poor credit ratings—even if the service had been perfect, collecting in the revenues would probably have been difficult.

iii.   *No evolution path for the technology*. BT's copper-based access network can be upgraded to broadband using ADSL technology. Amazingly, Ionica's radio-based technology had no upgrade path to broadband.

11. We saw in the previous chapter how the search for more capacity on trunk routes drove the evolution from microwave to optical systems. Transmitting a beam of light over long distances at ground level is not normally a viable option, but distances in the access network are usually short and a light beam can sometimes be transmitted for a few kilometers at ground level without the need for an optical fiber to protect it. Since light beams and microwaves are both forms of electromagnetic radiation, it is not surprising that these "free space optical" systems share many of the advantages and disadvantages of microwave radio. However, they are rather vulnerable to interference when used at ground level, so they tend to be confined to specialist applications. They have, for example, been used by Formula 1 racing teams for communication with track-side areas that are inaccessible to cabling. In this application, the portability of the equipment, speed of installation and the fact that it does not require a license are all key advantages. Free space optical systems have also been used to provide connections in countries where interference from bad weather is less of an issue and where the risk of copper wire being dug up and sold for scrap is very real.

12. The purpose of the conditioning units was to control electrical noise and to keep the impedance of the network within strictly defined limits. Although the mains-borne interference generated by modern domestic appliances is controlled by law, old and faulty equipment can generate enough interference to make powerline communications difficult or impossible. Furthermore, the impedance presented to the network by an individual household can vary from very low (when everything is switched on) to very high (when everything is switched off), so the propagation and attenuation characteristics of the network are subject to large and unpredictable fluctuations.

13. Powerline technology has been used successfully to create high-speed in-home networks that can compete with technologies such as WiFi.

### Chapter 8: Computers get chatty

1. The Xerox Palo Alto Research Center in California.
2. Coaxial cable was discussed in Chapter 6.
3. To be strictly accurate, the message would be discarded by the network

interface hardware. The computer would only be interrupted if a valid message arrived.

4. Computers on the Aloha radio network were allowed to transmit whenever they had data to send. They would then wait for confirmation from the destination that the packet had arrived. If no confirmation was received after a certain time, they would re-transmit the packet. No attempt was made to avoid collisions during transmission.

5. A switch that simply divided a LAN into two segments was sometimes referred to as a "bridge". Bridges tended to be implemented in software, so operated more slowly than hardware-based switches.

6. The switch can discover this information by looking at the source addresses of the messages on each segment.

7. Metcalfe's original network operated at 2.94 Mbit/second, but IEEE 802.3 Ethernet standards now exist for operation at 10, 100, 1,000, 10,000 and 100,000 Mbit/second. Most built-in laptop interfaces implement the 10/100 BaseT Ethernet standard, which can operate at 10 or 100 Mbit/second over twisted pair cable. High-speed router interfaces typically operate at 1,000 or 10,000 Mbit/second over fiber.

8. Wireless LANs have been standardized by the IEEE under such catchy titles as 802.11b and 802.11g. There are now a number of these standards using different radio bands, bit rates, etc., but they are collectively referred to as "WiFi".

9. Interference between wireless LANs should result in a reduction in network performance rather than any loss of data. The problem has been greatly reduced by the use of spread spectrum radio techniques.

10. Open Systems Interconnection.

11. A checksum can be used to check for errors in a received block of data. Before the block of data is transmitted, the checksum is calculated by performing some simple arithmetic or logical operation on the data. The checksum is then transmitted with the data. The operation is repeated at the receiver and the result is compared with the checksum sent by the transmitter. If the two checksums do not match, then the data has become corrupted during transmission. As a simple example, a checksum could be calculated for the data sequence 3, 9, 2, 7, 6 by adding the numbers together and using the least significant digit of the result. In this particular case, the digits add to 27, so the checksum would be 7.

12. To be strictly accurate, X.25 is not a single protocol, but a suite of compatible protocols. For example, the X.25 suite includes a Layer 3 protocol called PLP, a Layer 2 protocol called LAPB, and it can operate over a whole range of Layer 1 standards. Similarly, the AppleTalk suite contains components covering all seven layers in the ISO model.

13. As they are in the case of the X.25 protocol suite.

14. To illustrate just how dumb the network can be, an experiment was carried out in Norway to demonstrate that TCP could run over a network of carrier pigeons. The results of this experiment were published on April 1, 1990 as

"RFC 1149: A Standard for the Transmission of IP Datagrams on Avian Carriers". As the document states: "Avian carriers can provide high delay, low throughput, and low altitude service. The connection topology is limited to a single point-to-point path for each carrier, used with standard carriers, but many carriers can be used without significant interference with each other, outside of early spring. This is because of the 3D ether space available to the carriers, in contrast to the 1D ether used by IEEE802.3. The carriers have an intrinsic collision avoidance system, which increases availability." Further details, including photographs, can be found at *www.blug.linux.no/rfc1149/*.

## Chapter 9: The birth of the Internet

1. High-altitude nuclear explosions cause strong disturbances in the ionosphere, leading to radio wave scintillation. This can severely degrade radio and satellite communications over a wide area.
2. The name stands for "R&D".
3. Baran, P.: *On Distributed Communications Networks*, IEEE Trans. Comm. Systems, March 1964. See also RAND Paper P-2626 (*www.rand.org/pubs/papers/P2626*), which was written in 1962 and describes a general architecture for a large-scale survivable communications network.
4. Baran claimed that his inspiration for this concept came from Claude Shannon's mechanical mouse that could find its way through a maze (see Chapter 5).
5. Although the networks of the time were analog, Baran realized that a packet switching network would have to be digital if it was going to carry voice over long distances. Analog signals would be significantly degraded by a sequence of store-and-forward operations, while digital signals would be unaffected.
6. *www.lessig.org/content/standard/0,1902,7430,00.html*.
7. For example, they had chosen the same packet size and the same data transmission rate and both had developed adaptive routing algorithms.
8. Licklider, J. C. R. and Clark, W.: *On-Line Man Computer Communication*, August 1962.
9. Other predictions included point-and-click user interfaces, e-commerce and online banking.
10. ARPA changed its name to the Defense Advanced Research Projects Agency (DARPA) in 1971, then back to ARPA in 1993, and then back to DARPA in 1996. To avoid confusion, we will refer to them as ARPA throughout this book.
11. Some of these computers had specialized capabilities. For example, UCLA had focused on simulation, Utah on graphics and SRI on databases.
12. *http://inventors.about.com/library/inventors/bl_Charles_Herzfeld.htm*.
13. An interview with Lawrence G. Roberts, OH159, conducted by Arthur L. Norberg, April 4, 1989, Charles Babbage Institute, Center for the History of

Information Processing, University of Minnesota, Minneapolis (*www.cbi.umn.edu/oh/pdf.phtml?id=233*).

14. Roberts, L.: *Multiple Computer Networks and Intercomputer Communication*, ACM Symposium on Operating System Principles, Gatlinburg, October 1967.

15. Quoted in Hafner, K. and Lyon, M.: *Where Wizards Stay Up Late*, Touchstone Books, New York, 1996.

16. Their local Senator, Edward Kennedy, sent a slightly confused telegram to BBN congratulating them on their million-dollar contract to build the "Interfaith" Message Processor and thanking them for their ecumenical efforts! (*www.zakon.org/robert/Internet/timeline/*).

17. Leonard Kleinrock's Network Measurement Center at UCLA was selected to be the first node on the ARPANET.

18. In later years, BBN also developed Terminal IMPs (or TIPs). These allowed "dumb" terminals to be connected directly to the network, rather than having to connect via a host computer.

19. Purists would argue that the ARPANET protocols do not map exactly on to the standard OSI model. This is hardly surprising because the ARPANET protocols were developed before the model. In spite of this, it is useful to use the terminology of the model when discussing aspects of the ARPANET protocols.

20. Quoted in Reynolds, J. and Postel, J.: *RFC 1000—Request for Comments Reference Guide*, p. 3.

21. In previous experiments, host computers had been persuaded to talk to each other by programming one of them to behave like a peripheral device. The same "master–slave" form of communication was used on the ARPANET prior to the development of the Network Control Program. The problem with this form of communication was that only one of the hosts could issue instructions. A new protocol would be required to enable hosts to communicate in an equal, or "peer-to-peer", relationship.

22. Quoted in Reynolds, J. and Postel, J.: *RFC 1000—Request for Comments Reference Guide*, p. 3.

23. Later renamed the Network Control *Protocol*.

24. It was during the development of the Network Control Protocol that the Request For Comments (RFC) system was first established. RFCs were initially intended as a way of sharing ideas with other network researchers, but they subsequently developed into a key feature of the Internet standard-setting process. As Steve Crocker later recalled, the Network Working Group consisted mainly of graduate students, who naturally assumed that a team of experienced engineers would eventually turn up to take charge of the development. They therefore expressed their notes with some care:

> "I remember having great fear that we would offend whomever the official protocol designers were, and I spent a sleepless night composing humble words for our notes. The basic ground rules were that anyone could say anything and that nothing was official. And to emphasize the point, I labelled the notes 'Request for Comments'."

RFC1 was issued on April 7, 1969. It was soon found that the publication of one RFC would trigger other RFCs in response, thereby providing an open forum for technical discussions within the developer community. Once a consensus had been reached on a particular issue, a specification document would be prepared and the various research teams would then set about the task of implementing that specification. The RFC process developed into an effective method of collective decision making and is still in use today.

25. Leiner, Barry M., Cerf, Vinton G., Clark, David D., Kahn, Robert E., Kleinrock, Leonard, Lynch, Daniel C., Postel, Jon, Roberts, Larry G. and Wolff, Stephen: "A Brief History of the Internet", Internet Society, *www.isoc.org/Internet/history/brief.shtml.*

26. Cerf, V. G. and Kahn, R. E.: "A Protocol for Packet Network Interconnection", *IEEE Transactions on Communications*, **COM-22**(5), 627–641 (May 1974).

27. Appendix H describes some of the basic principles behind the funding of the modern Internet.

28. Vinton G. Cerf (left) and Robert E. Kahn (middle) received the Presidential Medal of Freedom from US President George W. Bush on November 9, 2005. Cerf and Kahn were honored for their work in helping to create the modern Internet.

## Chapter 10: Life in cyberspace

1. Quoted at *www.greatachievements.org/?id=3741.*

2. Although Tomlinson was unaware of it, @ was already in use on other systems as an escape character, a prompt or a command indicator. It was not until the 1980s that the role of @ in email addresses was finally accepted as a worldwide standard.

3. Berners-Lee, T.: *Weaving the Web*, Orion Business Books, London, 1999, p. 38.

4. Berners-Lee, T.: *Weaving the Web*, Orion Business Books, London, 1999, Foreword.

5. For users of Microsoft Internet Explorer v6.0, select View followed by Source.

6. Hafner, K. and Lyon, M.: *Where Wizards Stay Up Late*, Touchstone, New York, 1996, p. 252.

7. In order to reduce the opportunities for deception and fraud, many well known organizations register their name in every applicable domain. For example, Microsoft owns "microsoft.com", "microsoft.co.uk", "microsoft.biz", "microsoft.tv" and many others.

8. It is possible for a number of different hostnames to share a single IP address and this might be useful if a single web server is hosting a number of different websites. Conversely, a single hostname might share a number of different IP addresses and this might be needed if a number of physically separate web servers are supporting a single, very busy website.

9. If the web server is the only computer using the domain abc.co.uk, then the "www" is not strictly necessary and some websites don't bother with it.
10. The number of devices connected to the Internet is now so huge that the world is literally running out of IP addresses. Appendix I discusses some of the ways in which this problem is being addressed.
11. Often referred to as a "root server".
12. Google's founders, Larry Page and Sergey Brin, published a paper about Google when it was still an academic project at Stanford University. It can be found at *http://infolab.stanford.edu/~backrub/google.html*.
13. For example, you could not use a bulletin board to log on to a remote computer.
14. The choice of names is rather unfortunate. Usenet is not really a network and newsgroups are not news reports.
15. Initially, there were seven major categories of group (computing, news about Usenet, recreational activities, science, social issues, controversial issues and miscellaneous), with each category containing a number of groups and sub-groups catering to specific interests within that category. Since that time, Usenet newsgroups have proliferated to take in every conceivable area of human activity. Many are extremely specialist and some are frankly bizarre.
16. Berners-Lee, T.: *Weaving the Web*, Orion Business Books, London, 1999, p. 36.
17. Many instant messaging services provide a telephony option.
18. The initials RSS initially stood for RDF Site Summary. This was subsequently changed to Rich Site Summary and, today, RSS is generally considered to stand for Really Simple Syndication.
19. Each radio or TV program is downloaded as a file. Since the file is encrypted, it doesn't matter if it is copied thousands of times by a peer-to-peer file-sharing network. When someone wants to gain access to the contents of the file, a notification is sent automatically to a licensing server that is controlled by the owner of the content (i.e. the BBC). If the user is asking to do something legitimate, then a decryption key is returned along with details of what the user is allowed to do (view between certain dates, store, copy, print, etc.).
20. Some people have argued that grid computing is not a true peer-to-peer application because each of the participating computers is working selflessly for the good of the group rather than simply meeting its own needs. However, it can also be argued that a computer that uploads 100 copies of a file while participating in a file-sharing network is being equally selfless.
21. The performance had reached 528 TeraFLOPS. One TeraFLOP is 1,000,000,000,000 Floating Point Operations Per Second.
22. As we learned in Chapter 5, the quality of a digital signal is unaffected by the number of times that it is transmitted and received. An analog signal, on the other hand, is slightly degraded by every operation that is performed on it and the damage cannot be removed by subsequent processing.
23. A number of these devices could share a single broadband connection, thereby providing a cost-effective way of delivering multiple phone lines.

24. "Pixel" is short for "picture element".
25. Reviews of many different virtual worlds can be found at *www.virtualworlds-review.com*.
26. As with real-world currencies, the value of the Linden Dollar rises and falls in response to market pressures.

## Chapter 11: The mobile revolution

1. © Motorola, Inc., Heritage Services & Archives. Reproduced with permission.
2. Meurling, John and Jeans, Richard: *The Mobile Phone Book: The Invention of the Mobile Phone Industry*, Communications Week International, London, on behalf of Ericsson Radio Systems, 1994, p. 43.
3. Two-way radio communication was not established until 1933.
4. A channel is actually a pair of frequencies—one for each direction of transmission. However, echo problems on early mobile systems meant that full duplex conversation was not possible (see Appendix D for an explanation of echo). As in the case of a walkie-talkie, each user pressed a handset button to talk and then released the button to listen.
5. Using higher frequencies can also help to reduce the distance over which radio waves will propagate. The early radio telephone networks used frequencies that propagated much too well.
6. Cell coverage patterns are not really hexagonal (!), but this representation is useful for illustrating how they fit together.
7. There were some notable exceptions. The Dutch built the world's first nationwide public radio telephone network as early as 1949. By 1956, Sweden had mobile networks in Stockholm and Gothenburg, and Norway built its first radio telephone network in 1967 (Farley, Tom: "Mobile Telephone History", *Telektronikk*, **3/4**, 22–34 (2005)).
8. A single radio telephone call required as much spectrum as a broadcast radio station.
9. By 1976, only 545 customers in New York City had Bell System mobiles, but there were 3,700 customers on the waiting list ( Gibson, Stephen W.: *Cellular Mobile Radiotelephones*, Prentice Hall, Englewood Cliff, 1987, p. 8).
10. O'Brien, James: *Final Tests Begin for Mobile Telephone System*, Bell Laboratories Record, July/August 1978, p. 171.
11. It would be another 4 years before the FCC would finally authorize commercial cellular service in the United States. Although the Americans had pioneered the cellular concept, the development of commercial services was impeded by the lethargy of the FCC.
12. GSM originally stood for Groupe Speciale Mobile, after the study group that proposed the standard. It has since been re-named Global System for Mobile communications.
13. It is sometimes claimed that standardization stifles innovation. GSM

addressed this problem by specifying how the major components of the system should interwork, while leaving manufacturers free to be innovative in the design of the components.

14. A codec (short for coder–decoder) is used to convert analog voice to digital and vice-versa. For a discussion on voice codecs, see Chapter 5.

15. Called RPE–LPC. It stands for Regular Pulse Excited–Linear Predictive Coder, in case you were wondering.

16. Half Rate coding schemes enable two voice channels to be fitted into the bandwidth previously required for one and provide a convenient solution for operators with network capacity problems. Further bit rate reductions are possible by using voice activity detection to turn off the transmitter when the user is not talking. However, speech quality has now become a significant competitive issue and this has driven the network operators in a different direction. Enhanced full rate codecs still operate at 13 Kbit/second but the additional speech processing is used to improve speech quality rather than to reduce the bit rate.

17. The simplest way to design a mobile network would be to put an omni-directional BTS in the center of each cell. However, it turns out to be more efficient to locate each BTS at the point at which several cells meet. The BTS would be configured to transmit in three different directions on three different sets of frequencies, thereby effectively creating three separate cells. This technique is known as sectorization. The triangular arrangement of antennas at the top of a mobile mast reveals that it is being used in this way.

18. The dividing line between microcells and picocells is not well defined. Some people consider that picocells are used indoors while microcells are used outside.

19. It appears that the conversation took place on New Year's Eve, 1989. The tapes were published by *The Sun* in 1992.

20. *http://en.wikipedia.org/wiki/Squidgygate*.

21. Snooping software had been illegally installed on the systems of Vodafone Greece and conversations had been relayed to a recording system via pay-as-you-go mobile phones (*www.theregister.co.uk*, February 6, 2006).

22. A group of channels on the "air interface" between the BTS and the mobile phones are assigned to carry data. The ratio of data channels to voice channels on this interface can vary over time, depending on demand. The data channels are grouped together to form a single high-speed channel, with all the data traffic for the cell being carried on this channel.

23. The new modulation and channel coding schemes used by EGPRS are known as EDGE (Enhanced Data for Global Evolution). Despite the name, EDGE can be used to increase the capacity available for circuit switched voice services as well as for packet switched data services. The introduction of EDGE requires no hardware or software changes in the core network, but the base stations need to be modified and new mobile handsets are needed.

24. GPRS is referred to as a "2.5G" technology.

25. This technology, which is very different from anything used in GSM, is described in Appendix N.
26. This time, W-CDMA will be replaced by OFDMA (Orthogonal Frequency Division Multiple Access).
27. See Chapter 6 for a discussion on geostationary satellites.
28. Typically about 1,100 kilometers or less above the Earth's surface.

## Chapter 12: When failure is not an option

1. Appendix P provides some additional detail.
2. The decision to switch a sub-system out of service and replace it with a back-up sub-system has to be made automatically, which means that control software is required somewhere else in the system. There are several ways in which this decision can be made. Each sub-system will run a set of diagnostic software routines and will report in sick if any of these routines detects a problem. However, a fault in a sub-system could disrupt the operation of the diagnostic software. To detect this, the control software periodically sends a message to each sub-system and it expects to receive a reply—if one sub-system fails to respond, then it is assumed to be dead and the back-up sub-system is instructed to take over. If the in-service sub-system keeps reporting problems, then the control software may decide to switch to the back-up to see whether it can do any better.
3. As early as the 1960s, the AT&T 1ESS switch had a goal of no more than 2 hours' downtime in 40 years (Rennals, D. A.: *Fault Tolerant Computing, http://ftp.cs.ucla.edu/~rennels/article98.pdf*, p. 5).
4. It is, in fact, possible to reduce the time required for Network Restoration by carrying out the calculations in a distributed rather than a centralized way. By breaking up the problem and assigning it to a number of different devices in the network to work on in parallel, results can be obtained far more quickly. However, doing this sacrifices some of the most attractive features of Network Restoration because each device has only a limited view of what is going on. This means that the result is not optimized for the network as a whole and the response of the network to a particular type of failure becomes much harder to predict.
5. Both dedicated protection and network restoration depend upon the existence of redundant capacity in the network. Dedicated protection schemes require a dedicated back-up link for every active link in the network so the network operator has to build twice as much network capacity as will be required for normal operation. Network restoration, on the other hand, only requires about 30% overcapacity to allow the network operator to maintain service after a failure. By reducing the need for redundant capacity in the network, network restoration can deliver significant cost savings.
6. Some network operators use both dedicated protection and network restoration in an attempt to get the best of both worlds. Dedicated

protection is provided for high-value traffic to ensure that the impact of a failure is negligible. Network restoration is then used to provide a new back-up path for the dedicated protection scheme so that it can still respond if a second failure occurs. The network restoration also provides slower—but more cost-effective—resilience for less valuable traffic. So long as some spare capacity remains in the network, network restoration can protect against any additional failures that occur.

7. Appendix Q explains how errors in packets are detected. Chapter 8 discusses how Transport-layer protocols (such as TCP) can arrange for damaged or lost packets to be re-transmitted.

8. This raises an interesting question: if a transmission link between the routers is using dedicated protection or network restoration, then is it the transmission network or the routers that is responsible for putting things right after a network failure? The short answer is that resilience schemes in different layers of the network can work together very effectively but they can also conflict with each other. As we saw in the case of networks that use dedicated protection and network restoration, the aim is to exploit the best features of each approach. If dedicated protection is used on a link between two routers, then it will respond far faster than the routers to a break in the cable. For this reason, the routers would be configured to wait until the dedicated protection has had long enough to operate before taking any action. If both the primary and the back-up routes are cut, then the dedicated protection will be unable to fix the problem and the routers will fix it instead.

## Chapter 13: What comes next?

1. This has not happened since the early years of the telegraph.

2. To be strictly accurate, the call can be made anywhere in the vicinity of their home. The user registers their home address with the network operator and all calls using a cell that provides coverage at that address will be at fixed network rates.

3. "Smartphones", which provide an enhanced range of computing and data capabilities, are expected to represent about one-third of handset sales by 2013.

4. Streaming and downloading are two different methods of providing access to content stored on a network. A streamed television program would be viewed as the bits arrive (as on a conventional digital television), while a downloaded program would be stored and then watched at a later date (as on a VCR). Podcasting is a form of downloading.

5. DAB-IP is a mobile TV standard that is based upon an enhancement of the DAB digital radio network, so it can operate in frequency bands already allocated to digital radio. The DVB-H standard, on the other hand, requires a new transmitter network to be built and spectrum may not be allocated to

the service until after the switch-off of analog terrestrial television. However, the EU invested in the development of DVB-H and is promoting its use across Europe.

6. Of course, congestion can occur on circuit-based networks, but the effect on individual users is different. In the case of a telephone network, calls that are in progress should be unaffected by congestion, but attempts to set up new calls may be blocked until some existing calls have finished.

7. The ARPANET was designed to facilitate remote log-in to large mainframe computers. The extraordinary success of network email had not been anticipated.

8. This is referred to in marketing circles as "the long tail".

9. Competitions now exist in which small teams of programmers compete to create the most innovative mashup in under 24 hours. This particular application was developed at the Mashed event in London in June 2008.

10. The efficiency improvements are actually better than these figures would suggest because some spectrum has been reallocated from television broadcasting to other purposes.

11. Satellite TV has to use a dial-up modem link or a broadband connection to provide interactive services such as voting or home shopping. Does this matter? Possibly not, because the evidence so far seems to suggest that the public are fairly underwhelmed by most TV-based interactive services.

12. Digital TV supports two different types of interactivity. For fully interactive services such as audience voting or home shopping, two-way communication is required to allow the viewer to interact with the service. Cable TV can provide a suitable return path for this purpose, but terrestrial and satellite broadcasting are inherently uni-directional. This problem can be overcome by using a dial-up modem link or a broadband connection to provide the return path. However, some simpler interactive services (such as providing the recipe for a cookery program) do not need a return path; the additional information is broadcast alongside the program and the viewer can view this information if they wish by pressing the red button on their handset.

13. The technology behind IPTV is discussed in Appendix L.

14. At the time, most people connected to the Internet using dial-up access over BT telephone lines. However, Freeserve used telephone numbers belonging to Energis rather than BT, so Internet access calls to Freeserve had to be handed over from BT to Energis at the nearest point of interconnection between the two networks. BT would collect the full cost of the call from the customer, but a proportion of this revenue would then be handed over to Energis to pay them for terminating the call. A proportion of this payment was passed on to Freeserve to enable them to offer the "free" Internet access service.

15. This saying is usually attributed to John Wanamaker, a key figure in the history of advertising.

16. See Chapter 4.

17. Sandham, D.: "Fragile Web", *Engineering & Technology*, November 22–December 5, p. 70 (2008).

18. British computer hacker Gary McKinnon managed to break into sensitive American computer systems belonging to NASA and the US military. He claimed that he was searching for evidence of UFOs and "free energy technology" that was being deliberately covered up by the authorities.
19. Chapter 6 contains details of the electromagnetic spectrum.
20. A video that appeared on the Internet showing popcorn being cooked by placing it near mobile phones did nothing to allay public concerns, but it was subsequently shown to be a spoof that had been produced as part of a viral marketing campaign. The transmissions from mobile phones are, in fact, far too weak to cook popcorn—or even to make it slightly warm (*www.koreus.com/video/telephone-portable-mais-popcorn.html*).
21. The Mobile Telecommunications and Health Research Programme report for 2007 (*www.mthr.org.uk/documents/MTHR_report_2007.pdf*) provides a useful review of the evidence.
22. The speed of light causes a number of problems in telecoms networks. Although most of these can be hidden from users, some fiber optic links can generate long enough delays to affect high-speed data applications. The quarter-second delay introduced by satellite links is something that we see every day on our television screens when journalists are reporting from remote locations. Communication with probes in the outer reaches of the solar system is plagued by round-trip delays of up to a day and there is simply no prospect of holding a telephone conversation with any extra-terrestrial civilization that may be orbiting a distant star. Unless our current understanding of relativity theory turns out to be incorrect, the speed of light is one restriction that we are just going to have to live with.

## Appendix C: Microphone wars

1. William Orton (the head of Western Union) described the young Thomas Edison as a man with "a vacuum where his conscience ought to be".

## Appendix D: Digital signal processing

1. Quantization noise is explained in Chapter 5.
2. Eight bits can be arranged in 256 different ways (00000000, 00000001, 00000010 through to 11111111). This means that 8 bits can be used to represent 256 different voltage levels for each sample.
3. We learned in Chapter 3 how the young Alexander Graham Bell, working with his older brother, created a model representing all the main organs involved in human speech. The tongue was made of several small coated paddles, while the larynx came from a dissected sheep and a set of bellows provided the lungs. This apparatus apparently produced surprisingly human-like sounds. Source coders obviously use mathematical models rather than bits of sheep, but the principle is essentially the same.

4. This is not a new idea. A technique called Time-Assignment Speech Interpolation (TASI) was developed as early as 1959 to increase the capacity of analog submarine cables by switching the available capacity to those users that were actually speaking at any particular moment. This idea subsequently developed into Digital Speech Interpolation (DSI), which was used to increase the capacity of digital trunk circuits.

5. 1/10 of a second.

6. In practice, of course, connections across the core of the network will normally use multiplexers and fiber optic cables rather than copper wires, but it is helpful to think of each multiplexer channel as a separate pair of wires.

7. Amazing as it may seem, the speed of light is actually too slow for long-distance telecommunications. Light travels at about 200,000 kilometers/ second in an optical fiber (about two-thirds of its speed in a vacuum), so it takes about 30 milliseconds to travel from London to New York. This means that any echo will return after about 60 milliseconds, and this delay is quite long enough to make conversation difficult if echo cancellation is not used. The situation is even worse with communications satellites in geostationary orbits, because these introduce a round trip delay of over half a second.

## Appendix E: DSL technologies

1. See Chapter 4 for an explanation of Frequency Division Multiplexing.

2. Sometimes called a DSL filter. In the United Kingdom, it takes the form of a little white box that plugs into the telephone socket.

## Appendix F: Leveling up the playing field

1. We are talking about domestic and small business customers here. Operators are usually delighted to build dedicated fiber connections to large business customers if they generate lots of traffic.

## Appendix G: Fixed wireless access networks

1. To be strictly accurate, WiFi is normally used as a Local Area Network (LAN) rather than as a fixed wireless access technology. A conventional broadband connection is still needed to link the WiFi base station back to the network.

2. IEEE 802.16-2004.

3. IEEE 802.16e.

4. Many 3G network operators paid very high prices for their licenses at a time when mobile and the Internet were two of the most exciting technologies around and a network that brought these two technologies together was

seen as a marriage made in heaven. However, the competition from faster and cheaper (although much less mobile) fixed wireless access is stealing traffic from the mobile network operators in key locations such as airports and railway stations. The mobile operators, meanwhile, have the rather onerous obligation of providing coverage over large areas of the country where the revenue opportunities are likely to be far lower. We have become used to mobile networks stealing traffic from their fixed-network competitors, but when viewed from this perspective, the battle between fixed and mobile looks rather less one-sided.

5. Test results reported by Pipex in August 2006 found that an indoor aerial with no clear line of sight to the base station could deliver 2 Mbit/second in both directions over a distance of 1.2 kilometers. Using an external aerial with line of sight enabled 6 Mbit/second (downstream) and 4 Mbit/second (upstream) to be delivered over a distance of 6 kilometers. WiMAX is certainly capable of much longer ranges than this—particularly when it is used in licensed radio bands. However, longer ranges may not be desirable on heavily loaded networks. Since every user in the coverage area of a WiMAX base station is sharing the same pool of bandwidth, congestion can become a problem if the coverage is extended to take in too many users.

6. IEEE 802.11s.

## Appendix I: The Internet address shortage

1. The Internet Assigned Numbers Authority assigns blocks of addresses to five Regional Internet Registries. RIPE NCC is the Regional Internet Registry for Europe, the Middle East and parts of Central Asia (*www.ripe.net*).

2. Known as IP Version 6, or IPv6.

3. Since all packets arriving from the Internet carry the address of the router, how does the router know which computer on the home network to send them to? The answer lies in the use of additional bits in the packet header, but the details are beyond the scope of this book. The interested reader is referred to articles on network address translation that are available on the Internet.

## Appendix J: Virtual private networks

1. Some corporate VPNs use rather more sophisticated techniques to allow them to carry voice and video as well as data traffic.

2. The computers on the office networks would almost certainly be using "private" network addresses. These are IP addresses that cannot be used on the Internet because they have been assigned to another user, but they can be used within a private network (such as an office network or a home

network) where the rest of the world cannot see them. If any of these computers wish to access a site on the Internet, then network address translation (NAT) is used to replace the private address with a legitimate Internet address. However, NAT is not needed when sending packets over the VPN tunnel from one office network to another because the private IP addresses are hidden within the encrypted payload. The issues associated with private network addresses and NAT are discussed in Appendix I.

3. Firewalls can implement a wide range of security rules. For example, they can be configured to block particular protocols or particular IP addresses and they can filter out packets containing particular words or phrases. Firewalls can be configured to allow those forms of traffic that the business specifically needs (like email) but to block everything else.

## Appendix K: Internet voice services

1. Public Switched Telephone Network.
2. These services typically use the UDP protocol instead of TCP.

## Appendix M: GSM networks

1. Where a network operator has subsidized the cost of a mobile handset, there is always the risk that users will sign up for a subscription simply to obtain a cheap handset. To prevent this from happening, some operators use a technique called SIM locking to prevent their handsets from being used with other operators' SIM cards. This is illegal in some countries and has led to the appearance of handset unlocking services that will remove the lock in return for a fee.
2. If both cells are controlled by the same BSC, then a handoff between the cells can be managed by the BSC. If, however, the source and destination cells are controlled by different BSCs, then the handoff is managed at the MSC level. Things become even more complicated if the two cells are in areas controlled by different MSCs!
3. Since the comparison between the two keys has to be done over a radio link, there is an obvious risk of eavesdroppers. An ingenious technique is used that allows the comparison to take place without the need to transmit the secret key itself.
4. General Packet Radio Services.
5. The role of the GGSN in the packet switched network is similar to that of the Gateway MSC (GMSC) in the circuit switched network.

## Appendix N: Wideband CDMA

1. Check the Glossary if you are feeling a bit rusty about Time Division Multiplexing (TDM) and Frequency Division Multiplexing (FDM).
2. There are 124 carrier frequencies available in the 900 MHz band.
3. Or 16 voice channels if half-rate coding is used.
4. Wideband Code Division Multiple Access.
5. W-CDMA is normally used for 3G networks in countries that adopted GSM for their 2G networks. However, even countries that followed other standards are likely to use a modulation scheme based on the same principles.

## Appendix O: Network reliability

1. For critical software, it is possible to use separate development teams to create different versions of the same software. Each version of the software would run on a different operating system, so the chances of the same software bug appearing in more than one version of the software are very small. The results produced by each software version would be compared to determine whether an error has occurred and which version of the software is likely to be at fault. This technique is extremely expensive and is not normally justified for telecoms applications.
2. When one node crashed, it (correctly) sent "Out of Service" messages to its neighboring nodes. These messages told the neighboring nodes to route telephone calls around the failed node. However, the software bug caused the neighboring nodes to crash when they received the "Out of Service" message so that they, in turn, sent "Out of Service" messages to their neighboring nodes. Since all the nodes were running the same software (and so contained the same bug), the problem propagated rapidly through the network.

## Appendix P: Availability

1. In telecoms jargon, this is referred to as "five nines" availability.
2. For important customers, network operators will sometimes offer service level agreements that guarantee the availability of their services. Customers are entitled to receive service credits if a service fails to meet its agreed availability target.
3. In the access network, service availability normally depends upon network reliability and short repair times rather than resilience. One of the techniques used to improve access network reliability is to minimize the need for manual intervention. Experience shows that any work carried out on the access network tends to increase the risk of new faults!

## Appendix Q: Error detection and correction

1. In some systems, the rule is that the number of ones must be even. So long as there is consistency between the transmitter and the receiver, either odd or even parity will work.
2. To be strictly accurate, the Internet carries out error detection at several different levels. Each Ethernet link between routers has a checksum procedure of its own that protects not only the IP packet's header, but also its data. The TCP protocol adds a further checksum of its own. To avoid this duplication of error detection, the latest version of IP (IPv6) omits the checksum from the packet header.
3. If the errors had occurred in the same bit in two separate values, then the code would not have been able to correct them.

# Glossary

Here are some widely used telecoms terms that may not be familiar to non-technical readers. Italic text used in a definition indicates a term that is itself the subject of a definition.

**2.5G**
Technologies such as *GPRS* that are used to enhance 2nd Generation mobile networks, but do not provide the full capabilities of *3G* mobile networks

**3G**
3rd Generation mobile networks

**Access network**
The access network provides links from individual customers to a *local exchange* or a "point of presence". In a traditional telephone network, the access network consists mainly of *twisted pair* copper cables. See also *regional network, core network*

**ADPCM**
Adaptive Differential Pulse Code Modulation. A speech coding technique used to reduce the bit rate of a voice channel from 64 to 32 Kbit/second or less with minimal loss of speech quality

**ADSL**
Asymmetric Digital Subscriber Line. A technology used to deliver broadband services over conventional *twisted pair* telephone lines. The term "asymmetric" refers to the fact that the bandwidth provided from the network to the user is considerably greater than the bandwidth provided in the opposite direction. This improves the performance of the service for web surfing—but not for file sharing and other activities requiring a fast uplink

**Analog**
See *Digital*

**ATM**
Asynchronous Transfer Mode. A technology standard intended to replace *circuit switching* as the basis for multimedia networks. Although ATM has been relatively successful, the technology is now in decline as a result of the growth in *IP*. Many of ATM's best features have reappeared in *MPLS*

**Attenuation**
Attenuation is the ratio between transmitted and received power and represents the loss of signal that occurs in a transmission line. Attenuation is normally measured in decibels (or dB for short)

| | |
|---|---|
| **Availability** | Availability expresses the percentage of time that a telecoms service or a piece of equipment is fully operational |
| **Bandwidth** | For analog systems, bandwidth measures the range of frequencies that can be handled by the system (e.g. the bandwidth of an analog voice channel normally extends from 300 Hz to 3.4 kHz). For digital systems, bandwidth represents the capacity of the system measured in bit/second |
| **Base station** | See *cellular networks* |
| **Bluetooth** | Bluetooth is a radio standard that allows devices such as laptop computers, printers, PDAs, mobile phones and digital cameras to communicate over short distances. The standard is named after Harald Bluetooth, a tenth-century king of Denmark |
| **Broadband** | In telecoms parlance, the terms "broadband" and "narrow-band" meant, respectively, high and low *bandwidth*. However, these were relative terms and the boundary between broadband and narrowband was never very well defined. More recently, the term "broadband" has been used to describe high-speed always-on Internet access and to differentiate it from the much slower *dial-up* form of access |
| **Browser** | Computer program used to explore the World Wide Web |
| **Cellular networks** | Cellular networks provide radio coverage over a geographical area by dividing it into a number of cells. The radio coverage in each cell is provided by a base station. The base stations in adjacent cells operate on different frequencies in order to avoid interference. The power transmitted by a base station is restricted to the level necessary to reach the boundaries of the cell and so the same frequency can be re-used in other non-adjacent cells. This means that cellular networks use radio spectrum very efficiently and the low transmitter power is a major advantage when communicating with battery-powered mobile devices |
| **Circuit switching** | Circuit switching is used in conventional telephone networks to set up and clear down calls. Circuit switching is characterized by the establishment of a fixed-bandwidth circuit across the network for the duration of a call. A fixed-bandwidth circuit is not well suited to carrying "bursty" data traffic and the network resources required to support the circuit are tied up for the duration of a call, even if no useful information is being transmitted. For these reasons, networks of the future will use *packet switching* rather than circuit switching |

**Client–server**     In a client–server architecture, the client is frequently a PC or laptop computer on which the user is running one or more applications. *Servers* are available on the network to provide specific services to the client. These services might include file storage, database access and printer management. In the context of the Internet, servers are used to host websites, while the client PC runs the web browser

**Coaxial cable**     A form of copper cable in which two concentric conductors are separated by an insulator (or "dielectric"). Coaxial cable delivers much better performance than other forms of metallic cable

**Conditional access**     A security system used to protect broadcast content from unauthorized access. In the case of satellite television, anyone with a suitable dish can receive the signal, but a conditional access system ensures that people who have not paid for the service cannot view the television channel. Conditional access works by encrypting the signal before it is broadcast and then supplying authorized viewers with the necessary decryption key (usually in the form of a card that is plugged into the satellite decoder)

**Core network**     Most large networks can be divided into a core (or backbone) network and a number of *regional networks* and *access networks*. The core network provides the high-capacity infrastructure that links together large towns and cities

**Dial-up**     Connecting to another computer using a modem and a conventional telephone call (as opposed to *broadband*, which provides an always-on connection)

**Digital**     *Analog* signals can be transmitted by networks in either analog or digital format. An analog representation of a signal uses a continuously variable parameter (such as voltage) to track variations in the signal. In contrast to this, digital signals use a sequence of (usually binary) digits to represent the values of successive signal samples. An analog representation of a signal can hold any value within a given range, while digital signals can only hold a restricted set of values as a result of *quantization*. The early telephone networks were entirely analog, but digital technology offers major advantages

**Digital Rights Management**     Any form of content (music, video, text, etc.) can be digitally encoded and stored in a file on a computer. This makes it very easy for fraudsters to make illegal copies—unlike analog, digital content can be copied any number of times with no loss of quality. Digital Rights Management

(DRM) provides a mechanism for content owners to protect their intellectual property. The content is stored in encrypted form and cannot be viewed or listened to without the decryption key. This key is obtained automatically by the content player from a license server on the Internet. However, the key will only be issued if the user is attempting to use the content in a valid way. For example, DRM can be used to restrict the number of times that a movie can be viewed or the number of different music players that a song can be stored on

**Domain Name System (DNS)**   The system that translates an alpha-numeric domain name (such as *www.abc.co.uk*) into an IP address

**DSLAM**   Digital Subscriber Line Access Multiplexer. If *broadband* services are delivered using *ADSL* over *twisted pair* telephone lines, a DSLAM is required at the *local exchange* to terminate the copper pairs

**Duplex (or full duplex)**   Full duplex communication can take place simultaneously in both directions, as in the case of a telephone call. See also *simplex*

**DWDM**   Dense Wavelength Division Multiplexing. A form of *FDM* that can be used to dramatically increase the available capacity of an optical fiber. Instead of transmitting just one beam of light, a large number of optical signals are transmitted at different frequencies on the same fiber

**EDGE**   Enhanced Data for Global Evolution. New modulation and channel coding schemes that can be used to increase radio link capacity in GSM networks. The introduction of EDGE requires no hardware or software changes in the core network, but the base stations need to be modified and new mobile handsets are needed

**Ethernet**   A popular Local Area Network technology. See *LAN*

**FDM**   Frequency Division Multiplexing. Multiplexing is used to combine a number of low-speed channels into a single high-speed aggregate. FDM is a method of transmitting multiple signals down a single cable or radio link by sending them at different frequencies. Analog radio and television broadcasts use FDM to keep the channels apart. See also *TDM*, *statistical multiplexing*

**Fiber optic**   A form of cabling in which signals are transmitted optically rather than electrically. The huge potential bandwidth available in fiber optic cables means that they are very widely used in *core* and *regional networks* and are becoming increasingly common in *access networks*

| | |
|---|---|
| **Firewall** | A security device to protect private networks from certain types of Internet threat |
| **Frequency** | Frequency is a measurement of the number of cycles per second completed by an oscillating waveform (such as a radio wave). The Hertz (Hz) is the standard measure of frequency and was named in honor of Heinrich Hertz. 1 Hz is defined as one cycle per second, but for most telecoms applications, we need to work in kHz (thousands of cycles per second), MHz (millions of cycles per second) or GHz (thousands of millions of cycles per second). There is a very close relationship between *wavelength*, frequency and speed of propagation. For light or radio waves, Wavelength × Frequency = Speed of Light |
| **Frequency Division Multiplexing** | See *FDM* |
| **FTP** | File Transfer Protocol. A standard method of transferring files from one computer to another |
| **Full duplex** | See *duplex* |
| **Geostationary** | Satellites orbiting the equator at 35,786 kilometers above mean sea level have the property that they remain in a fixed position in the sky when viewed from the Earth. Satellites in this orbit are referred to as "geostationary". Although these satellites introduce a noticeable transmission delay because of the distances involved, they are widely used for satellite broadcasting and for other forms of communication where fixed (non-steerable) satellite dishes must be used |
| **GPRS** | Early GSM networks could support a dial-up data channel running at 9.6 Kbit/second—the equivalent of a rather slow modem link. This was adequate for exchanging small amounts of data with people on the move, but it was totally unsuitable for more demanding applications. To address this issue, mobile networks built a data capability in parallel with their existing voice capability. In the case of GSM networks, the new data capability was called General Packet Radio Services—or GPRS for short |
| **GSM** | A "2nd Generation" mobile network technology. GSM originally stood for Groupe Speciale Mobile, after the study group that proposed the standard. It has since been renamed Global System for Mobile communications |
| **Half duplex** | See *simplex* |

**HLR**
Home Location Register. This is the database in a GSM network where customer details are stored. Every active *SIM card* issued by the mobile network operator has its own HLR entry. This entry holds information such as the SIM's unique identifier, the corresponding mobile telephone number(s), the services that the subscriber is permitted to use, configuration options and information relating to the subscriber's current location. This is pretty important information. If the HLR dies, the network dies with it

**Host**
A computer on a network that provides facilities to remote users. For example, if you access a website, the information is being downloaded to you by a network host

**HTML**
HyperText Mark-up Language is a language used to create web pages. If you view a web page using Microsoft Internet Explorer, you can see what HTML looks like by selecting View/Source

**HTTP**
HyperText Transfer Protocol is used on the World Wide Web to transfer information between clients and servers. See *client–server*

**IP**
Internet Protocol. IP is the network protocol on which the Internet (and many other packet networks) are based

**ISDN**
Integrated Services Digital Network. A circuit-switched network providing an end-to-end digital connection. Voice traffic is normally transmitted at 64 Kbits/second, while data might be carried at higher bit rates. ISDN connections were widely used in some places during the 1990s, but they are now in decline because they offer less bandwidth than packet switched networks. The need to set up and clear down a call every time data needs to be transferred makes ISDN much less suitable than a packet switched network for carrying bursty data

**ISP**
Internet Service Provider. Retail ISPs sell Internet access services to domestic customers or businesses. Tier 1 ISPs provide the *core network* for the Internet and are often used by retail ISPs to provide connectivity to other parts of the Internet

**LAN**
Local Area Network. A network that is (typically) restricted to a single building. See also *MAN, WAN*

**Leased line**
A digital connection between two sites operating at a fixed bit rate. Unlike *ISDN*, a leased line is an "always-on" connection, so it does not need to be set up or cleared down

**LEO**            Low Earth Orbit satellites are much closer to the Earth than
                   *geostationary* satellites. This has advantages and disadvan-
                   tages. LEO satellites introduce a much shorter transmission
                   delay than geostationary satellites, but they are harder to use
                   for communication purposes because they do not remain in
                   a fixed location in the sky and regularly disappear below the
                   horizon. To provide continuous coverage, a "constellation"
                   of LEO satellites is required

**Local call**     Local calls are typically calls between two subscribers
                   connected to the same local exchange

**Local exchange** The cable that connects each customer to the network
                   terminates at the local exchange. To a telephone engineer,
                   the term "telephone exchange" refers to the building and all
                   the equipment inside it. The piece of equipment that
                   physically routes telephone calls across a network is
                   generally referred to simply as a "switch"

**Local loop**     In most countries, telephone networks were originally built
**unbundling**     by nationalized monopolies such as BT, France Telecom and
                   Deutsche Telekom. As telecoms markets around the world
                   were liberalized, these companies evolved into dominant
                   players in competitive markets. Since control of the copper
                   telephone lines that connect customers to the network is a
                   major strategic advantage, regulators forced them to allow
                   competitors to rent copper lines and to install equipment in
                   local exchanges. Local loop unbundling is used by network
                   operators to offer high-speed broadband services over
                   copper cables that do not belong to them

**Long-distance**  In simple terms, a long-distance call is any call that is not a
**call**           *local call*. In practice, the situation is often more complicated
                   than this; calls between adjacent exchanges are sometimes
                   treated as local calls, while some operators have drawn a
                   distinction between long-distance calls and regional calls

**MAN**            Metropolitan Area Network. A network that is restricted to a
                   single town or city. See also *LAN, WAN*

**Microwave**      Microwaves are electromagnetic waves with frequencies in
                   the range 1–300 GHz (1 GHz is $10^9$ Hz). This means that
                   they fit between radio waves and infrared in the electro-
                   magnetic spectrum. Microwaves are used for communicat-
                   ing with satellites and microwave-based *transmission* systems
                   can provide a convenient way of implementing network
                   *trunks* over difficult terrain. Since each microwave transmit-
                   ter must have a clear line of sight to its receiver, microwave
                   towers are often built on top of hills or tall buildings

| | |
|---|---|
| **Modem** | Modem is short for MOdulator/DEModulator. The word "modem" is used in a number of different contexts, but it is typically a device that enables digital data to be transmitted over an analog telephone network |
| **MPLS** | Multi Protocol Label Switching. Packet-based networks can introduce unnecessary delays because a separate routing calculation is carried out in every router that a packet passes through. If a packet requires a major long-distance route (such as New York to Los Angeles), it should be possible to put the packet into a "pipe" in New York and have it re-emerge at Los Angeles without the need for routing decisions at intermediate nodes. MPLS is often implemented in core network routers, where it is used to improve traffic management in IP networks. However, the technology turns out to have a number of other benefits. It can, for example, be used to construct *VPN*s or to separate traffic with different Quality of Service requirements |
| **Multiplexing** | Multiplexing is used to combine a number of low-speed channels into a single high-speed aggregate. Available multiplexing techniques include *FDM*, *TDM* and *statistical multiplexing* |
| **Narrowband** | See *broadband* |
| **Packet switching** | In packet switched networks, data are transmitted in discrete lumps called packets. Each packet carries a header, which provides important information about the packet (such as the source and destination address). Next-generation networks based on packet switching have started to displace traditional networks based on *circuit switching* |
| **PCM** | Pulse Code Modulation. The standard digital encoding technique used in telephone networks. Each telephone channel requires a bit rate of 64 Kbits/second |
| **PDH** | Plesiochronous Digital Hierarchy. A standard form of *Time Division Multiplexing* used in early digital networks. Now largely displaced by *SDH* |
| **Peer-to-peer** | In a *client–server* computer network, the role of the clients is different from that of the servers. In contrast to this, all computers in a peer-to-peer network are treated as equals, or peers. Peer-to-peer file-sharing networks are often (sometimes illegally) used for the distribution of music or video content. Skype uses peer-to-peer technology to carry telephony traffic across the Internet |
| **POTS** | Plain Old Telephone Service. Refers to the basic telephony service offered by traditional fixed telephone networks and |

is sometimes used to differentiate these services from mobile telephony and *VoIP*. The term PANS (Pretty Advanced Network Services) was suggested to cover enhancements to POTS such as voicemail, conference calling and caller ID, but the term has never really caught on

**Private circuit**   See *leased line*

**Protection**   A range of techniques used to improve network *availability* through the use of *redundancy*. Protection techniques typically require two separate cable routes and service is maintained so long as at least one of the cable routes remains intact. Protection techniques operate faster than network *restoration* techniques, but make less efficient use of bandwidth

**PSTN**   Public Switched Telephone Network. This term relates to the traditional telephone network that we all know and love. It does not include Internet-based telephony

**Quantization**   Quantization is an operation that occurs during the process of converting an *analog* signal to *digital*. The value of each sample is measured and it is then converted to the nearest available digital equivalent. This process of forcing a measurement to adopt one of a restricted number of possible values is known as quantization. A quantizer with a large number of possible output values will provide an extremely accurate digital representation of an analog signal but a high bit rate will be required to transmit the digital information across the network

**Redundancy**   Most techniques designed to improve network *availability* are based, in one form or another, on the concept of redundancy. Redundant equipment is not required when the network is fully functional, but it is brought into service (usually automatically) when something fails

**Regional network**   Most large networks can be divided into a *core* (or backbone) *network* and a number of *regional networks* and *access networks*. A regional network, as its name suggests, provides the network infrastructure in a specific geographical region. The regional network sits between the low-speed infrastructure in the access network and the high-speed infrastructure in the core network

**Reliability**   A reliable network is one with a low level of faults. Network reliability would normally be achieved by high-quality engineering combined with effective operational procedures. However, even a highly reliable network might not be good enough for some services (e.g. those carrying high-value financial transactions). In these situations, investment is required in network *resilience*

**Resilience**

A resilient network is one that continues to deliver services to customers in spite of network faults. Techniques such as *protection* and *restoration* are used to improve network resilience

**Restoration**

A range of techniques used to improve network *availability* through the use of *redundancy*. Restoration techniques re-route services that have been affected by a node or link failure, using whatever spare bandwidth is available in the network. Restoration techniques operate more slowly than *protection* techniques, but make more efficient use of bandwidth

**Router**

Routers are the devices used at the nodes in *packet switching networks*. Packets are forwarded from one router to the next as they travel across the network. If part of a network fails, the remaining routers will attempt to re-route packets around the problem

**Satellite**

See *geostationary*, *LEO*

**SDH**

Synchronous Digital Hierarchy. A standard form of *time division multiplexing*. Supports much higher speeds than *PDH* and enables individual channels to be added or dropped without the need to de-multiplex the whole payload. SDH also incorporates a range of *protection* mechanisms. Although SDH is now regarded as a "legacy" technology, it is still widely used

**Server**

A computer that provides services to other computers. For example, a web server might host a number of websites, while a file server supports file downloads. Servers are often used in *client–server* architectures

**SIM card**

A Subscriber Identity Module (SIM) is a smart card used to store subscriber and network information in a GSM mobile phone. The stored information can include security codes, subscription details, saved telephone numbers, user preferences and text messages

**Simplex**

In this book, simplex communication is considered to be communication that is permanently restricted to just one direction of transmission (as in the case of television broadcasting), while half duplex communication can only operate in one direction at a time (as in the case of walkie-talkie radios). However, there has been some unfortunate inconsistency in how these terms are used, with some people claiming that walkie-talkie radios are actually an example of simplex communication

| | |
|---|---|
| **Statistical multiplexing** | Multiplexing is used to combine a number of low-speed channels into a single high-speed aggregate. Available multiplexing techniques include *FDM*, *TDM* and statistical multiplexing. In statistical multiplexing, it is assumed that the channels will require capacity at different times, so that it should be possible for them to share bandwidth on a single link without causing bits to be lost. The technique exploits the statistical properties of the data to reduce the bandwidth required. Statistical multiplexing is often implanted using *packet switching* |
| **Switch** | See *local exchange* |
| **Tandem exchange** | A telephone exchange that is used to route calls efficiently between local exchanges. To a telephone engineer, the term "telephone exchange" refers to the building and all the equipment inside it. The piece of equipment that physically routes telephone calls across a network is generally referred to simply as a "switch" |
| **TCP** | Transmission Control Protocol. Often written as TCP/IP because of TCP's close association with *IP*. TCP provides reliable communication over an IP network. See also *UDP* |
| **TDM** | Time Division Multiplexing. Multiplexing is used to combine a number of low-speed channels into a single high-speed aggregate. TDM is a method of transmitting multiple signals down a single cable or radio link by sending them at different moments in time. See also *FDM*, *statistical multiplexing* |
| **Time Division Multiplexing** | See *TDM* |
| **Transmission** | Transmission systems are used to implement network *trunks*. In *core networks*, transmission systems are typically based upon *fiber optics* or *microwaves* |
| **Triple play** | A bundle of services that is offered to customers as a package. It normally includes television, high-speed Internet access and a fixed telephone service. The term "quadruple play" has been used for packages that also include mobile services |
| **Trunk** | A high-capacity link between two network nodes |
| **Twisted pair** | A form of copper cable in which pairs of conductors are twisted together to reduce the effect of external interference. Twisted pair cable is often used for domestic telephone lines |
| **UDP** | User Datagram Protocol. An alternative to *TCP* that is used in situations in which delivering packets with minimum |

|  | delay is more important that ensuring that every packet is received uncorrupted. UDP is typically used for transmitting voice and video over the Internet |
|---|---|
| **UMTS** | Universal Mobile Telecommunications System. A 3rd Generation mobile network technology |
| **URL** | Uniform Resource Locator. The method used on the World Wide Web to specify the location of resources. A URL is the address that you type into your browser when you want to view a web page |
| **VLR** | Visitor Location Register. Each MSC in a *GSM* mobile network uses a VLR to hold information relating to the active subscribers in its area |
| **VoIP** | Voice over IP. A technology used to transmit voice traffic over an *IP* network by converting it into a packetized form. In some cases, packets are only generated when someone is actually speaking, thereby reducing the total number of packets required to transmit a telephone conversation |
| **VPN** | Virtual Private Network. A technology that uses capacity on a public network to provide private networking facilities for use by an individual organization. VPNs provide the security and predictability of a private network at lower cost |
| **WAN** | Wide Area Network. A network that stretches beyond the boundaries of a single town or city. See also *LAN*, *MAN* |
| **Wavelength** | The distance between two successive peaks of a waveform, measured in meters. There is a very close relationship between wavelength, *frequency* and speed of propagation. For light or radio waves, Wavelength × Frequency = Speed of Light |
| **WiFi** | A standards-based wireless LAN technology. See also *LAN* |
| **WiMAX** | A standards-based radio technology with applications in Wireless Local Loop (*WLL*) and 4th Generation mobile networks |
| **WLL** | Wireless Local Loop. A generic term covering radio-based technologies (such as WiMAX) used in the access network. WLL technologies are sometimes used instead of copper telephone lines |
| **X.25** | An early standard for connecting devices to packet switched networks |

# Bibliography

Arianrhod, Robyn: *Einsteins' Heroes*, Icon Books, Cambridge, 2004.

Bannister, Jeffrey, Mather, Paul and Coope, Sebastian: *Convergence Technologies for 3G Networks*, John Wiley & Sons, 2004.

Beauchamp, Ken: *History of Telegraphy: Its Technology and Applications*, Institution of Engineering & Technology, 2000.

Berners-Lee, Tim: *Weaving the Web*, Orion Business Books, London, 1999.

Bodanis, David: *Electric Universe: How Electricity Switched on the Modern World*, Little, Brown, London, 2005.

Calvert, J. B.: *The Electromagnetic Telegraph*, May 2004.

Casson, Herbert N.: *The History of the Telephone*, A. C. McClurg & Co, Chicago, 1910.

Dr K., *A Complete H@cker's Handbook*, Carlton Books, London, 2000.

Fares, Alex: *GSM Systems Engineering & Network Management*, Authorhouse, 2003.

Hafner, Katie and Lyon, Matthew: *Where Wizards Stay Up Late*, Touchstone, New York, 1996.

Huitema, Christian: *Routing in the Internet*, Prentice Hall, 1995.

Naughton, John: *A Brief History of the Future*, Weidenfeld & Nicolson, London, 1999.

Pope, Frank L.: *Electric Telegraph: A Handbook for Electricians & Operators*, D. Van Nostrand, New York, 1881.

Shannon, Claude E.: "A Mathematical Theory of Communication", *Bell System Technical Journal*, **27**, 379–423, 623–656 (July/October 1948).

Spufford, Francis: *Backroom Boys*, Faber & Faber, 2003.

Standage, Tom: *The Victorian Internet*, Phoenix, 2003.

Tisal, Joachim: *The GSM Network*, John Wiley & Sons, Chichester, 2001.

Vise, David A.: *The Google Story*, Pan Books, 2006.

Von Baeyer, Hans Christian: *Information: The New Language of Science*, Weidenfeld & Nicolson, London, 2003.

# Index